# The Book of Science Days

*An everlasting diary*

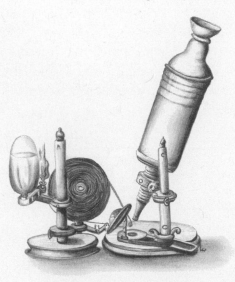

*Written by Anne Hance*

*Illustrations and design by Kathy West*

*First Edition published November 2000*

*Created on behalf of*
*Explorit Science Center*
*Davis, California*

*The Printer, Davis, California*

# Dedication

*Explorit Science Center's mission is:*
*"To involve people in science experiences that touch our lives."*

The Center works to encourage science literacy through its
particular style of science learning opportunities.
We dedicate this book to experiences that encourage
a child's natural curiosity, and sense of wonder to
be sustained into adulthood as driving forces
behind a lifelong fascination with science.

Published by Explorit Science Center
P.O. Box 1288, Davis, California 95617-1288, USA

First edition

ISBN 0-9705533-0-7

Printed in the United States of America

# Acknowledgements

Explorit Science Center would like to thank the authors of the essays, and the author of the introduction, for their contributions to the book. Thanks also go to the Cousteau Society for its permission to include the excerpt from Jacque Cousteau's 1984 letter to the Explorit Science Center.

Special thanks for substantial financial support to help pay our publishing costs are due to Novo Nordisk Biotech Inc. of Davis, California, an anonymous donor, and several individuals who made no-interest loans.

Thank you Anne for thinking of this project, nurturing its development, and meticulously researching and writing the book, Pearl for helping at the very beginning of the project, Sandy for being an interested and active provider of new sources of information, Evelyn for directing Anne to one of the main sources of information about actual birthdates which have often been quite hard to find, and very special thanks to artist Kathy West for her wonderful drawings and beautiful layout.

*Explorit Science Center*

# Table of Contents

# Preface

Identifying the birthday of a scientist for each day of the year, our everlasting diary provides a glimpse into the history of science. Who was adding to the store of scientific knowledge in the 1500s? Where in the world have the majority of scientists been born or studied?

Our earliest historical birthday (April 15, 1452) is the birthday of Leonardo da Vinci, the first birthday in the diary is that of Nicolaus Steno (January 1, 1638) and the last is the birthday of Andreas Vesalius (December 31, 1514). Between January 1 and December 31 the brief entries for each day describe studies and discoveries often revealing how individual scientists have influenced one another.

As you read the entries on the 366 days of the year, and perhaps assemble them in your head in chronological order, it should become apparent that the development of scientific knowledge is an uneven flow of small, incremental new and revised understandings punctuated with occasional swells which result from occasional great discoveries.

This Book of Science Days includes many more women scientists than most such lists. We would have included Hypatia, a Greek, who was perhaps the first woman scientist but, although her life and works are quite well documented, we could not determine the actual date of her birth in Alexandria in AD 370. Most of her writings have been lost but references to them exist and these tell us that she was a mathematician and a philosopher whose most significant work was in algebra but who was interested in mechanics and practical technology. She is said to have designed several scientific instruments including a plane astrolabe, an apparatus for distilling water, and a graduated brass hydrometer.

Our book has been created to further the mission of the nonprofit Explorit Science Center in Davis, California, U.S.A. which exists to foster interest in science and understanding of the ways in which science affects our lives.

We started collecting the data for the diary in 1990 and have been working on and off in our spare time for almost ten years. At first the material was in a large three-ring binder with 366 blank sheets of lined paper! Then, we moved to an Osborne 'Portable' Computer with a CPM operating system, a 10 MB hard drive, and a five-and-a-half inch floppy disk drive. When our project was being completed at the turn of the century we were using Apple laptops with gigabytes of hard disk space and cutting edge software. This book has seen science history in action!

We imagine that many people will use the diary simply to dip into every now and again, while others will use it as a book of birthdays and anniversaries in their own lives. We anticipate that it will be useful for teachers in stimulating their students to studying about scientists and their science. We hope you enjoy this Book of Science Days.

*Anne Hance*
*Explorit Science Center, July 2000*

# Introduction

One of the challenges a historian of science faces is to communicate the tremendous scope of scientific investigation, the range of its discoveries, and the diversity of its practitioners. This book of days goes a long way toward meeting this challenge as it moves from sixteenth century botanist Leonhart Fuchs to twentieth century astronomer Jocelyn Bell Burnell, from Chinese physicist Chien Shen Wu to Indian biochemist Har Khorana. The range of entries offered here, wonderfully demonstrates that great science has been practiced in a number of different ways by men and women from a variety of cultures and societies.

When Fuchs produced his beautifully illustrated herbals, he catalogued the wonders of the natural world available to him. I doubt he could have even imagined a world with Jocelyn Bell Burnell's pulsars or the nucleic acids of Har Khorana's biochemistry. Indeed, Fuchs worked at a time when the practice of botany looked back to ancient Greek traditions championed by Dioscorides and Theophrastus. While modern botanists may admire Fuch's herbals, they work in a time dominated by laboratory and experimental traditions that make botany today a radically different kind of science.

I hope that this book of days will help foster both a sense of the historical diversity of science and scientists, as well as an appreciation of the value that this diversity of opinion and practice has for science today and in the future.

*Michael R. Dietrich*
*History and Philosophy of Science*
*University of California, Davis*

*Shooting Stars*

# January

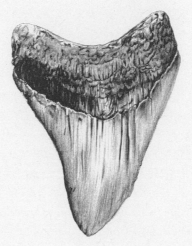

*"During the century after Newton, it was still possible for a man of unusual attainments to master all fields of scientific knowledge. But by 1800, this had become entirely impracticable."*

*Isaac Asimov (U.S. biochemist) January 2*

*Fossilized Shark's Tooth*

# Introduction to January

## *I Caught my Addiction from a Friend of my Father and a 200-Year-old Book*

The man from whom I caught the addiction to mathematics was a Hungarian University Professor and friend of my father, Dr. Klug. He gave me a 200-year-old book to read when I was eleven years old. The book was entitled "Algebra." It was written by the great mathematician Euler. It was simple and made sense. All it needed was the pleasure to understand something that is consistent and surprising.

My university studies were completed in Germany at the wonderful period when the surprising ideas of relativity and quantum mechanics were developed. This great intellectual development had been shortly afterwards crushed by Hitler.

I became professor of Physics at the George Washington University in Washington D.C. and had a wonderful time in applying the new ideas to old puzzles in chemistry and structure of matter. With the discovery of nuclear fission, the start of World War II and the possibility of release of this nuclear energy, I became deeply interested in nuclear bombs. My interest was initiated and sustained by association with contemporary Hungarian scientists, Eugene Wigner, Johnny von Neumann and Leo Slizard.

Applications of these new fields of knowledge remained at the center of my interests. I am still working on safe and reliable nuclear energy sources to provide us with electricity long after we have exhausted conventional sources such as coal and oil.

*Edward Teller*
*University of California, Berkeley*
*July 1997*

# January

| | | | | |
|---|---|---|---|---|
| 1 | Nicolaus Steno, 1638 - 1686 | | 18 | Edward Frankland, 1825 - 1899 |
| 2 | Isaac Asimov, 1920 - 1992 | | 19 | James Watt, 1736 - 1819 |
| 3 | Robert Whitehead, 1823 - 1905 | | 20 | Edwin E. (Buzz) Aldrin, 1930 - |
| 4 | Brian David Josephson, 1940 - | | 21 | Horace Wells, 1815 - 1848 |
| 5 | Joseph Erlanger, 1874 - 1965 | | 22 | Francis Bacon, 1561 - 1626 |
| 6 | Surveyor 7, 1968 | | | André Marie Ampère, 1775 - 1836 |
| 7 | Jacques-E. Montgolfier, 1745 - 1799 | | 23 | Paul Langevin, 1872 - 1946 |
| 8 | Alfred Russel Wallace, 1823 - 1913 | | | Gertrude B. Elion, 1918 - 1999 |
| | Stephen William Hawking, 1942 - | | 24 | Jacques de Vaucanson, 1709 - 1782 |
| 9 | Soren P. Lauritz Sorensen, 1868 - 1939 | | | Desmond John Morris, 1928 - |
| | Har Gobind Khorana, 1922 - | | 25 | Robert Boyle, 1627 - 1691 |
| 10 | Robert Woodrow Wilson, 1936 - | | 26 | Roy C. Andrews, 1884 - 1960 |
| 11 | William Curtis, 1746 - 1799 | | 27 | Dmitri I. Mendeleeff, 1834 - 1907 |
| 12 | Lazarro Spallanzani, 1729 - 1799 | | 28 | Kathleen Y. Lonsdale, 1903 - 1971 |
| 13 | Wilhelm Wien, 1864 - 1928 | | | Space Shuttle Challenger, 1986 |
| 14 | Matthew F. Maury, 1806 - 1873 | | 29 | Alice Evans, 1881 - 1975 |
| 15 | Edward Teller, 1908 - | | 30 | Max Theiler, 1899 - 1972 |
| 16 | Dian Fossey, 1932 - 1985 | | 31 | Irving Langmuir, 1881 - 1957 |
| 17 | Leonhart Fuchs, 1501- 1566 | | | Explorer I, 1958 |
| | Benjamin Franklin, 1706 - 1790 | | | |

# January 1

*Nicolaus Steno (Biology)*
*Birthday: 1638, Copenhagen, Denmark*

Nicholaus Steno, also known as Niels Steensen, studied in Copenhagen under anatomist Thomas Bartholin. Later, in Paris, he made important observations on the anatomy of the brain; and compared shark's teeth with certain inland fossils and deduced that certain inland areas had once been seas. At about this time, Robert Hooke (July 18) was considering similar ideas.

*Fossilized Shark's Tooth*

# January 2

*Isaac Asimov (Biochemistry)*
*Birthday: 1920, Petrovichi, Russia*

Isaac Asimov studied at Columbia and taught biochemistry at Boston from 1948-58, but is most renowned as a first rate, prolific writer of science fiction. His first science fiction story was sold to Amazing Stories magazine (1939) and his first book was *Pebble in the Sky* (1950). He authored a biochemistry textbook and wrote many books on science for the general public.

# January 3

*Robert Whitehead (Engineering)*
*Birthday: 1823, Bolton, Lancashire, England*

Robert Whitehead was apprenticed at age 14 as an engineer in Manchester. Later (1866) in Italy with his son John, he invented a torpedo capable of running 700 yards at 7 knots. By 1889 he had increased the range to 1000 yards at 29 knots by improvements including adding a servo motor to the steering mechanism. Seven years later, he added a gyroscope to improve the torpedo's accuracy.

# January 4

*Brian David Josephson (Physics)*
*Birthday: 1940, Cardiff, Wales*

Brian David Josephson studied physics at Cambridge and remained there as professor of physics at the Cavendish (October 10) Laboratory. He shared a 1973 Nobel Prize in Physics for work on super conductors and semiconductors and discovered the Josephson Effect which is applied in high speed switching devices for computers. He later developed an interest in phenomena of intelligence.

# January 5

*Joseph Erlanger (Biology)*
*Birthday: 1874, San Francisco, CA, USA*

Joseph Erlanger studied medicine at Johns Hopkins. Later, At Washington University in St Louis, he made observations about the function of nerve synapses, and the transmission of electrical impulses along nerve fibers and across synapses. With medical pharmacologist Herbert Gasser, he devised new ways to amplify and record electrical nerve impulses. They shared 1944 Nobel Prize for Physiology or Medicine.

# January 6

*Surveyor 7*
*Event: 1968, US Spacecraft Launch*

Surveyor 7, an unmanned lunar probe spacecraft, was launched on January 6, 1968 and landed on the moon near Crater Tycho four days later. It performed some soil analysis, sent back 20,993 images from the surface of the Moon, and made the first observation of artificial light from Earth.

# January 7

*Jacques-Etienne Montgolfier (Invention)*
*Birthday: 1745, Vidalon-les-Annonay, France*

Jacques-Etienne Montgolfier and his brother Joseph-Michel were pioneers of lighter-than-air flight. They raised their first hot-air balloon 70 ft on November 15, 1782; on June 5, 1873 they sent up a larger, paper-covered canvas sphere carrying 400 lbs of ballast; on November 21, 1873 two passengers rose 3,000 ft on a 7 1/2 mile, half-hour flight.

# January 8

*Alfred Russel Wallace (Natural History)*
*Birthday: 1823, Usk, Monmouthshire, England*

Alfred Russel Wallace had no college education. Encouraged by entomologist H.W. Bates, he became interested in natural history and spent 8 years in the Malay Archipelago collecting and observing. He wrote an essay on the formation of new species which caught the attention of Charles Darwin (February 12) and Wallace joined with Darwin in 1858 to publish their common belief about evolution - the theory of natural selection.

*Stephen William Hawking (Physics)*
*Birthday: 1942, Oxford, England*

Stephen Hawking graduated in physics from Oxford in 1962 and went to Cambridge to work on relativity theory. He investigated the properties of black holes and the behavior of matter in the immediate vicinity of black holes. Hawking authored a popular book about the origin and future of the universe, *A Brief History of Time* (1988), for the non-mathematical layman.

*Black Hole*

# January 9

*Soren Peter Lauritz Sorensen (Biochemistry)*
*Birthday: 1868, Havrebjerg, Denmark*

Soren Sorenson studied chemistry and medicine at the University of Copenhagen. He became interested in the synthesis of amino acids, but is best known for pioneering research on hydrogen ion concentration. He introduced pH as the measure of acidity and alkalinity that is used universally around the world by laymen as well as scientists.

*Har Gobind Khorana (Biochemistry)*
*Birthday: 1922, Raipur, India*

Har Khorana became interested in the biochemistry of nucleic acids as a fellow at Cambridge working under A.R. Todd. He shared the 1968 Nobel Prize in Physiology or Medicine (with Nirenberg and Holley) for his interpretation of the genetic code and its function in protein synthesis. In 1970 he synthesized a DNA gene of yeast and later of E. coli (*Escherichia coli*).

*E. coli*

# January 10

*Robert Woodrow Wilson (Astronomy)*
*Birthday: 1936, Houston, Texas, USA*

Robert Woodrow Wilson attended Rice University and Cal. Tech. Working with Arno Penzias to measure radio signals from galactic hydrogen he found unexplained excess noise that they eventually identified as radio remnants of the cosmic Big Bang origin of the universe. This discovery helped to persuade many astronomers that the Big Bang theory is correct. Wilson shared the 1978 Nobel Prize in Physics.

# January 11

*William Curtis (Botany)*
*Birthday: 1746, Alton, Hampshire, England*

William Curtis was apprenticed at age 14 to his grandfather, a surgeon. He set himself up as an apothecary in London, but sold his practice in 1771 to establish a botanic garden in Bermondsey. Later, he established one in Lambeth, and finally one in Brompton. In 1787, he founded *The Botanical Magazine*, an immediate success, that continues today.

# January 12

*Lazarro Spallanzani (Natural History)*
*Birthday: 1729, Scandiano, Italy*

Lazarro Spallanzani studied law at Bologna and accepted the chair in natural history at the University of Pavia. He was a founder of experimental biology, working mainly with amphibians. He studied sterilizing by heat, showed that hot air is less effective than hot water for sterilization purposes, and showed that some minute organisms can live for days in a vacuum.

# January 13

*Wilhelm Wien (Physics)*
*Birthday: 1864, Gaffken, East Prussia, Germany*

William Wien was awarded the 1911 Nobel Prize in Physics for his studies on thermal radiation, hydrodynamics and x-ray measurements. He studied the phenomena produced when electric discharges are passed through gases and showed that Canal Rays were positively charged particles deflected by electric and magnetic fields.

*"If I set out to prove something, I am no real scientist—I have to learn to follow where the facts lead me—I have to learn to whip my prejudices."*

*Lazzaro Spallanzanzi (Italian Naturalist) January 12*

# January 14

### Matthew Fontaine Maury (Oceanography)
### Birthday: 1806, Fredericksburg, Virginia, USA

Matthew Fontaine Maury entered the US Navy in 1825, but was lamed in 1839 and restricted to shore duty. As Superintendent of Charts and Instruments in Washington, he issued a unique Wind and Current Chart of the North Atlantic (1847), and later added one for the Indian Ocean. In 1853 he produced a *Physical Geography of the Sea* which was the first textbook of modern oceanography.

# January 15

### Edward Teller (Physics)
### Birthday: 1908, Budapest, Hungary

Edward Teller emigrated to the US in 1935 and became a naturalized citizen. As a theoretical physicist he was part of the 1943-5 team that developed the first fission device in the Manhattan Project (July 16) in Los Alamos, New Mexico. He then worked on hydrogen fusion. In later years he developed the Lawrence Livermore National Laboratory in California to provide a balanced, two-laboratory approach to the problems of nuclear defense.

*Mountain gorilla*

# January 16

### Dian Fossey (Anthropology)
### Birthday: 1932, San Francisco, California, USA

Dian Fossey met Louis Leakey (August 7) in 1963 and under his sponsorship went to the high rain forests of the volcanic mountains on the Rwanda-Zaire-Uganda border in Central Africa to study mountain gorillas. She concentrated on observing behavior and culture in natural surroundings (ethology) and was eventually accepted at gorilla family gatherings. She later focused on preservation. She wrote *Gorillas in the Mist* (1983).

# January 17

*Leonhart Fuchs (Botany)*
*Birthday: 1501, Wemding, Bavaria*

Leonhart Fuchs, with degrees in classics and in medicine, was professor of medicine at Tubingen for most of his life. While practicing medicine he also gained renown as a botanist and by 1542 had published a herbal describing about 400 German and 100 foreign plants. The plant and the color fuchsia are named after him.

*Benjamin Franklin (Physics)*
*Birthday: 1706, Boston, Massachusetts, USA*

Benjamin Franklin was instrumental, in 1743, in starting America's first scientific society - The American Philosophical Society - in Philadelphia. In June 1752 he performed his famous kite experiment and showed that lightning is a form of electricity. Franklin is considered one of the great founders of the science of electricity.

# January 18

*Edward Frankland (Chemistry)*
*Birthday: 1825, Churchtown, Lancaster, England*

Edward Frankland studied in London under Lyon Playfair (May 21). He worked under Robert Bunsen (March 31) and spent a year with von Liebig (May 12). He introduced the concept of valency, which he initially called atomicity. (Wichelhaus later coined the word valenz). Frankland's work with Lockyer on solar spectra led to the discovery of helium.

# January 19

*James Watt (Engineering)*
*Birthday: 1736, Greenock, Scotland*

James Watt, perhaps the greatest of British engineers, trained in London as a mathematical instrument maker. Asked in 1764 to repair a Newcomen (February 24) steam engine, he noted deficiencies in the design and developed an improved engine which was manufactured in partnership with Matthew Boulton (September 3) in 1776. He also devised a chemical method for copying documents.

# January 20

*Edwin E. (Buzz) Aldrin (Space Exploration)*
*Birthday: 1930, Montclair, New Jersey, USA*

Edwin (Buzz) Aldrin received his Ph.D. from MIT with a thesis on the factors involved in achieving rendezvous between two orbiting spacecraft. Aldrin was one of two men - Neil Armstrong (August 5) was the other - to be the first humans to set foot on the Moon on July 20, 1969. His first words as he stepped onto the lunar surface were: 'Beautiful . ... magnificent desolation.'

*Moonprint*

# January 21

*Horace Wells (Medicine)*
*Birthday: 1815, Hartford, Vermont, USA*

Horace Wells began studying dentistry at age 19 and went into practice in Boston.  In 1844, while watching a demonstration of laughing gas (nitrous oxide), he noticed one of the volunteers badly bruise his legs without being aware of the pain. He gave nitrous oxide to a friend for a tooth extraction and confirmed its use as an anesthetic. He continued to use it for extractions.

# January 22

*Francis Bacon (Philosophy)*
*Birthday: 1561, London, England*

Francis Bacon was obsessed by the belief that he was born to be of service to humanity by the discovery of truth. He was a politician and essayist and argued against magic. He promoted the development of the scientific method by which an exhaustive collection of instances (data) must be analyzed by induction in order to arrive at true conclusions.

*Francis Bacon*

*André Marie Ampère (Physics, Mathematics, Chemistry)*
*Birthday: 1775, Near Lyons, France*

André Ampère was a tutored and self-taught genius who founded the science of electromagnetics in the 1820s and devised Ampere's Law in 1827. His name is given to the Ampere, a unit of electrical current flow.

# January 23

*Paul Langevin (Physics)*
*Birthday: 1872, Paris, France*

Paul Langevin studied for a short time under J.J. Thomson (December 18) at the Cavendish (October 10) Laboratory in Cambridge. He then returned to Paris to work with Pierre Curie. In 1915 Langevin invented SONAR (sound navigation ranging) mainly as a method for ships to avoid icebergs, but also applicable as a submarine detector.

# January 23

*Gertrude B. Elion (Biochemistry)*
*Birthday: 1918, New York City, USA*

Gertrude Elion shared the 1988 Nobel prize in physiology or medicine. She is one of the few Nobel Prize winners without a doctorate degree. She developed treatments for gout and herpes, laid the way for the development of AZT (azidothymidine), and helped change the way new drugs are discovered.

---

# January 24

*Jacques de Vaucanson (Invention)*
*Birthday: 1709, Grenoble, France*

Jacques de Vaucanson studied mechanics in Paris and while there he made his first automation, a flute player. He designed various lathes and drills to enable him to create his inventions such as the automated flute player and a swimming, quacking duck. As an inspector of silk factories, he invented and made the first self-acting loom, and improved the Dutch ribbon loom.

*Desmond John Morris (Zoology, Ethology)*
*Birthday: 1928, Wiltshire, England*

Desmond Morris attended Birmingham and then Magdalen College Oxford, where he worked on animal behavior under Nicolaas Tinbergen (April 15). He was head of Grenada Television, and Curator of Mammals for the Zoological Society of London. A prolific writer, he is well known for his books, films and television programs on human and animal behavior. He achieved national attention with his book *The Naked Ape* (1967).

*Desmond Morris*

# January 25

*Robert Boyle (Physics)*
*Birthday: 1627, Lismore Castle, County Cork, Ireland*

Robert Boyle had a prodigious memory and a flair for languages. He was educated privately and at Eton, and read widely. His early experiments were based on the use of the air pump. The law now known as Boyle's Law regarding the behavior of gasses was in fact discovered in 1662 by one of his assistants (R. Towneley).

# January 26

*Roy Chapman Andrews (Zoology)*
*Birthday: 1884, Beloit, Wisconsin, USA*

Roy Chapman Andrews was a member of the scientific staff of the American Museum of Natural History, New York. He led an expedition to Mongolia to search for the origin of man. Andrews wrote *On the Trail of Ancient Man* describing his experiences in the Gobi desert where in fact he found no human remains, but did find the first fossilized dinosaur eggs known to science.

# January 27

*Dmitri Ivanovich Mendeleeff (Chemistry)*
*Birthday: 1834, Tobolsk, Siberia*

Dmitri Mendeleeff studied chemistry at St Petersburg, Paris and Heidelberg. His great contribution to chemistry was the Periodic Table of Elements. For his book, *Principles of Chemistry* (1869), he developed the Periodic Law — a system of classifying the 60 then known elements and in which elements arranged in order of atomic weights were found to be grouped in useful ways.

# January 28

*Kathleen Yardley Lonsdale (Crystallography)*
*Birthday: 1903, Newbridge, Ireland*

Kathleen Yardley Lonsdale grew up in the suburbs of London, England, studied mathematics and physics at Bedford College, and did most of her groundbreaking work at Leeds University. Her discovery, published in 1929, that the benzene ring of organic molecules is a flat hexagon had a great effect on organic chemistry of the time. Through her work calculating tables of X-ray patterns in common crystals Lonsdale became the first woman crystallographer to gain an international reputation.

*Space Shuttle: Challenger (Space Exploration)*
*Event: 1986, US Space Shuttle Launch*

The Challenger, a US space shuttle, launched from Cape Kennedy (Canaveral) USA on January 28, 1986. It blew apart 72 seconds later, killing six astronauts and a civilian passenger. The cause of the explosion was traced to a leak in the solid booster rocket.

# January 29

*Alice Evans (Biology)*
*Birthday: 1881, Neath, Pennsylvania, USA*

Alice Evans obtained degrees in bacteriology from Cornell and Madison. Employed in 1910 by the US Dept. of Agriculture, Dairy Division, she contracted undulant fever and is best known for clarifying the role of pasteurization of milk in killing *Bacillus abortus* which causes undulant fever. In 1928 she was the first woman president of the Society of American Bacteriology.

*Alice Evans*

# January 30

*Max Theiler (Medicine)*
*Birthday: 1899, Pretoria, South Africa*

Max Theiler was a South African microbiologist who completed medical training at St. Thomas Hospital and the London School of Tropical Medicine. In 1937 he introduced a vaccine against yellow fever. He also discovered a natural virus disease in mice which is very similar to human poliomyelitis. He won the 1951 Nobel Prize for Medicine or Physiology.

# January 31

*Irving Langmuir (Chemistry)*
*Birthday: 1881, Brooklyn, New York City, USA*

Irving Langmuir obtained his doctorate from the University of Gottingen and spent most of his career working for General Electric Company in Schnectady. He was awarded the 1932 Nobel Prize in Chemistry for his work on surface chemistry - the properties of liquid surfaces. Langmuir worked during WW II on the production of smokescreens and he later applied this knowledge to the use of solid carbon dioxide and silver iodide in cloud seeding. (See also November 13).

*Explorer I*
*Event: 1958, US Space Satellite Launch*

Explorer I was the first US Earth satellite. The 6-foot, 31 pound, cylindrical Explorer I was launched on a Jupiter C rocket and reached an apogee of 1,587 miles and perigee of 212 miles. Explorer 1 discovered a radiation belt around the Earth, which was later identified by James Van Allen of the University of Iowa.

*Explorer I*

*"Men love to wander, and that is the seed of our science."*

*Ralph Waldo Emerson (American philosopher) 1803-1882*

# February

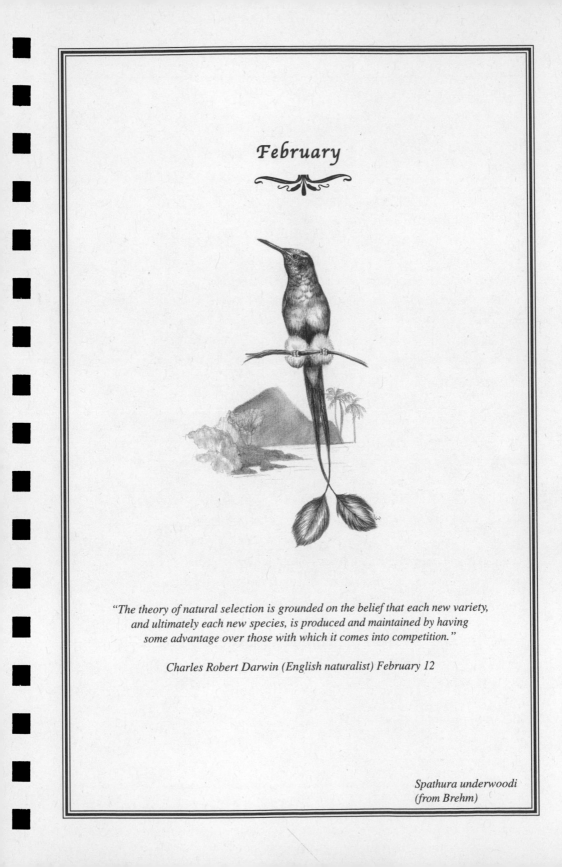

*"The theory of natural selection is grounded on the belief that each new variety, and ultimately each new species, is produced and maintained by having some advantage over those with which it comes into competition."*

*Charles Robert Darwin (English naturalist) February 12*

*Spathura underwoodi*
*(from Brehm)*

# Introduction to February

## The Essence of Science

I have always been curious about the world. New ideas and new experiences make me wonder, first about the unfamiliar, then about the things we take for granted. The essence of science for me is making observations, recognizing an unsolved problem, asking the right question, seeking answers with new evidence, and putting it all together in a coherent framework. The intense beauty of the objects and ideas of the world and the power of science as a way of knowing provide the satisfaction that continues to draw me to the pursuit of knowledge about things I do not understand.

My introduction to science came through an early and continuing appreciation of shells. Seen as abstract works of art, they provide an astonishing diversity of elegantly shaped and sculptured variations on the simple yet powerful theme of the logarithmic spiral. As diaries, they record the conditions and events of their makers' lives today and in the distant past. As pieces of functional architecture, shells are houses, whose features give us a glimpse into the factors that shaped the evolution of the animals that built the shells and of the creatures that lived with them. From their study, perhaps we can learn how to keep our own house in order.

*Geerat Vermeij*
*Dept. of Geology, University of California, Davis*
*May 1997*

*Recipient: MacArthur Foundation genius award*
*Author: A Natural History of Shells (1993); Privileged Hands (1996)*

# February

| | | | | |
|---|---|---|---|---|
| 1 | Emilio Segre, 1905 - 1989 | | 16 | John Rex Whinefield, 1901 - 1966 |
| 2 | Charles Manning Child, 1869 - 1954 | | 17 | Thomas Robert Malthus, 1766 - 1834 |
| | Salyut 4, 1977 | | 18 | Galileo Galilei, 1564 - 1642 |
| 3 | Elizabeth Blackwell, 1821 - 1910 | | | Alessandro Volta, 1745 - 1827 |
| 4 | Charles A. Lindbergh, 1902 - 1974 | | 19 | Nicolas Copernicus, 1473 - 1543 |
| | Clyde William Tombaugh, 1906 - 1997 | | 20 | Ludvig Boltzmann , 1844 - 1906 |
| 5 | Robert Hofstadter, 1915 - | | | Mercury 6 Capsule Friendship 7, 1962 |
| 6 | Eduard Hitzig, 1838 - 1904 | | 21 | John Mercer, 1791 - 1866 |
| | Mary Douglas Leakey, 1913 1996 | | 22 | Heinrich Rudolph Hertz, 1857 - 1894 |
| 7 | Ulf Svante Von Euler, 1905 - | | 23 | Allen McCleod Cormack, 1924 - 1998 |
| 8 | Daniel Bernoulli, 1700 - 1782 | | 24 | Thomas Newcomen, 1663 - 1729 |
| 9 | Howard Taylor Ricketts, 1871 - 1910 | | 25 | Phoebus A. Theodor Levene, 1869 - 1940 |
| 10 | Victor Hensen, 1835 - 1924 | | 26 | Dominique Arago, 1786 - 1853 |
| 11 | Thomas Alva Edison, 1847 - 1931 | | 27 | Petrus Artedi, 1705 - 1735 |
| 12 | Charles Robert Darwin, 1809 - 1882 | | | Alice Hamilton, 1869 - 1970 |
| 13 | William Bradford Shockley, 1910 - 1989 | | 28 | Rene A. F. de Reaumur, 1683 - 1757 |
| 14 | Julius Arthur  Nieuwland, 1878 - 1936 | | | Linus Pauling, 1901 - 1994 |
| 15 | Cyrus Hall McCormick, 1809 - 1884 | | | Peter Brian Medawar, 1915 - 1987 |
| | | | 29 | Charles Pritchard, 1808 - 1893 |

# February 1

*Emilio Segre (Nuclear Physics)*
*Birthday: 1905, Tivoli, Italy*

Emilio Segre was the first graduate student of Enrico Fermi (September 29) and focused his study on atomic spectroscopy. In 1934 he joined Fermi's group to do neutron research. He discovered technetium, the first artificial element, and astatine; and was awarded the 1959 Nobel Prize in Physics for the discovery of the anti-proton.

# February 2

*Charles Manning Child (Zoology)*
*Birthday: 1869, Ypsilanti, Michigan, USA*

Charles Manning Child received his doctorate from Leipzig University and spent most of his academic career at the University of Chicago. Child's main interest was embryology and he did significant work with various invertebrate animals tracing the way in which individual cells of their embryos developed.

*Salyut 4*
*Event: 1977 Soviet Space Station Launch*

Salyut 4, the Soviet Union's first civilian space station had been launched December 26, 1974. It had two successive cosmonaut crews who worked in the station January 11 - February 9, 1975, and May 25 - July 26, 1975. The spacecraft's orbit slowly decayed and it re-entered Earth's atmosphere and burned up on February 2, 1977.

# February 3

## Elizabeth Blackwell (Medicine)
### Birthday: 1821, Bristol, England

Elizabeth Blackwell emigrated with her family to the U.S. in 1832. She qualified from Geneva Medical School in New York, becoming the first woman doctor in the U.S., and set up private practice in that city. Patients were slow to come to a woman doctor and in 1853 she opened a one-room dispensary. This expanded and in 1857 became the New York Infirmary for women and children.

*Elizabeth Blackwell*

# February 4

## Charles A. Lindbergh (Aviation)
### Birthday: 1902, Detroit, Michigan, USA

Charles Lindbergh was an aviator and in 1927 made the first nonstop, solo flight across the Atlantic Ocean in 33.5 hours. Lindbergh also associated with scientists at the Rockefeller Institute and assisted Alexis Carrel, a French surgeon at the Institute, in the design of a sterilizing glass pump to circulate nutrient fluid. Lindbergh and Carrel jointly published *The Culture of Organs.*

## Clyde William Tombaugh (Astronomy)
### Birthday: 1906, Streator, Illinois, USA

Clyde William Tombaugh was an amateur astronomer when he discovered the planet Pluto on February 18, 1930 at Lowell Observatory in Flagstaff, Arizona. An employee at the observatory, he was working on the search for the ninth planet instigated by astronomer Percival Lowell in 1905. After his discovery, Tombaugh attended the University of Kansas to earn a degree and then returned to the Lowell Observatory cataloging more than 30,000 celestial objects before he left in 1946.

# February 5

*Robert Hofstadter (Physics)*
*Birthday: 1915, New York, USA*

Robert Hofstadter discovered that protons and neutrons have a definite structure. By passing a beam of electrons close to each type of particle and studying the patterns that resulted from the electrons' deflection, he determined that each particle had a positively charged central core surrounded by two shells. He was awarded a 1961 Nobel Prize in Physics.

# February 6

*Eduard Hitzig (Medicine)*
*Birthday: 1838, Berlin, Germany*

Eduard Hitzig was professor of psychiatry and director of the mental asylum in Zurich and later at Nietleben. In his researches he investigated the electrical excitability of the cerebral cortex of the dog. His goal was a more scientific approach to the treatment of mental patients and he established a neuro-psychiatric clinic in Halle.

*Mary Douglas Leakey (Paleoanthropology)*
*Birthday: 1913, London, England*

Mary Leakey made important discoveries in East Africa which helped revise theories about where the human species first appeared. In 1948 she unearthed the 18 million year old *Proconsul africanus* at Lake Victoria in Kenya. In 1959 she discovered the *Zinjanthropus* (now *Australopithecus*) *boisei* jaw at Olduvai Gorge. She worked in Africa with her husband Louis Leakey.

# February 7

*Ulf Svante Von Euler (Biology)*
*Birthday: 1905, Stockholm, Sweden*

Ulf Svante Von Euler shared the 1970 Nobel Prize for Physiology or Medicine with Julius Axelrod (May 30) and Sir Bernard Katz (March 26) for illuminating the chemical nature of nerve transmission. He showed that it is norepinephrin rather than epinephrin that is the transmitter substance of the sympathetic nervous system.

# February 8

*Daniel Bernoulli (Mathematics)*
*Birthday: 1700, Groningen, Netherlands*

Daniel Bernouilli, one of the best known of the famous Bernouilli family, received his doctorate in medicine for a thesis on the action of the lungs, but is best known as a mathematician for his important contributions to hydrodynamics and differential equations. His major published work, *Hydrodynamica* (1738) develops his subject (using Newton's laws of force) and includes many practical applications of his theories.

*Daniel Bernoulli*

# February 9

*Howard Taylor Ricketts (Medicine)*
*Birthday: 1871, Findlay, Ohio, USA*

Howard Taylor Ricketts studied at Nebraska and at Northwestern Universities. He discovered that the vector for Rocky Mountain spotted fever is a tick and that only adult ticks transmit the disease to humans. In Mexico he studied the disease tabardillo and showed it to be typhus fever and to be transmitted by a body louse.

# February 10

*Victor Hensen (Biology)*
*Birthday: 1835, Schleswig, Germany*

Victor Hensen studied science and medicine at Kiel and graduated in 1859. He became chair there in physiology and worked in embryology, histology and marine biology. He published significant work in connection with liver function, sense of touch, and hearing. Hensen gave the name plankton to the populations of drifting plant and animal organisms of the ocean.

# February 11

*Thomas Alva Edison (Invention)*
*Birthday: 1847, Milan, Ohio, USA*

Thomas Alva Edison had little formal education and during the Civil War was a telegraph operator. Later, he invented the gramophone, a long-lasting carbon-thread electric lamp, and the nickel cadmium accumulator. His 1,097 patented inventions were the result of hard work, trial and error. His inventions were said by the U.S. Congress to have revolutionized civilization.

# February 12

*Charles Robert Darwin (Natural History)*
*Birthday: 1809, Shrewsbury, Shropshire, England*

Charles Robert Darwin studied medicine at Edinburgh but left after two years and went briefly to Cambridge. He first made his name as a geologist but after nearly five years sailing round the world as naturalist on the H.M.S. Beagle his major contribution to science was his theory that evolution occurs as a result of natural selection of heritable traits. His theories and explanations were described in the *Origin of Species* (1859) and The *Descent of Man* (1871). Ironically, Darwin was developing his ideas while Gregor Mendel (July 22) was doing his revealing work on breeding pea plants — which was not published until 1900.

*Spathura underwoodi*
*(from Brehm)*

# February 13

*William Bradford Shockley (Physics)*
*Birthday: 1910, London, England*

William Shockley grew up in California, graduated from MIT and then worked at the Murray Hill Bell Telephone Laboratories in New Jersey in a semiconductor research group planning to replace vacuum tubes. With Walter Brattain and John Bardeen (May 23), he invented the point contact transistor. He then designed a junction transistor and shared the 1956 Nobel Prize for Physics with his collaborators.

# February 14

*Julius Arthur Nieuwland (Chemistry)*
*Birthday: 1878, Hansbeke, Belgium*

Julius Nieuwland grew up in Indiana and studied at Notre Dame. He taught botany there for 14 years, but is famous for his work in acetylene chemistry showing that acetylene can be converted to divinyl-acetylene, then polymerized to a rubbery material. He worked with DuPont to develop neoprene in 1931 - the first commercially successful synthetic rubber.

# February 15

*Cyrus Hall McCormick (Engineering)*
*Birthday: 1809, Walnut Grove, Virginia, USA*

Cyrus McCormick, an agricultural engineer, improved upon his father's unsuccessful attempt to design an effective reaper. He developed and manufactured a mechanical reaper that used a knife blade and cutter bar and was pulled rather than pushed by a horse. He obtained the patent for his invention in 1834 but did not manage to manufacture it successfully until the mid 1840s.

# February 16

*John Rex Whinefield (Chemistry)*
*Birthday: 1901, Sutton, Surrey, England*

John Whinefield read chemistry at Cambridge and assisted Charles Cross (December 11) who had invented rayon in 1892. Whinefield went to work for the Calico Printers Association and, in 1941, with J.T. Dickson, he invented Terylene which was marketed in 1947 by Imperial Chemical Industries (ICI) in Britain, and in the U.S. by DuPont as Dacron.

# February 17

*Thomas Robert Malthus (Economics)*
*Birthday: 1766, Dorking, Surrey, England*

Thomas Malthus was professor of political economy at Haileybury College. He is best known for his theories on population science and economics. In his *Essay on the Principle of Population* (1798, 1803), Malthus argued that populations have a natural tendency to increase faster than the needed food supply. This attracted attention even though the same ideas had been advanced a long time before by Pliny, Aristotle and others.

# February 18

*Galileo Galilei (Physics)*
*Birthday: 1564, Pisa, Italy*

Galileo was a founder of classical physics and a vigorous, creative scientist. He was a prominent astronomer and used the newly invented telescope to discover the mountains on the Moon, the nature of the Milky Way, and four of Jupiter's moons. He openly expressed his agreement with the Copernican theory that Earth is not the center of the universe. He developed some basic physical proofs including the theory of parabolic ballistics, and laws of fall, which were to be cornerstones of the work of Huygens (April 14) and Newton (December 25) who developed classical physics a generation later.

*Galileo Galilei*

# February 18

### Alessandro Volta (Invention)
### Birthday: 1745, Como, Italy

Alessandro Volta was professor of Physics at several universities, including Pavia, where he was made Rector. His invention of the electric battery (voltaic cell) changed the science of electricity. He disproved Luigi Galvani's (September 9) thesis that electrical currents can be generated by animal tissue. In 1799 he produced electricity in the absence of animal matter when he constructed the first chemical battery in which he used silver and zinc disks piled alternately with absorbent material soaked in water between each disc. He did not, however, understand that a chemical reaction produced the electricity.

# February 19

### Nicolas Copernicus (Astronomy)
### Birthday: 1473, Thorn, Poland

Copernicus was a canonicus of the Cathedral of Frauenburg for the whole of his adult life. He studied the classics and medicine, observed and recorded astronomical events, and developed a theory that the center of Earth is the center of gravity and of the lunar orbit but not of the universe. He believed that the apparent movement of the stars was actually due to the movement of Earth. His new ideas were heretical according to the Church of the day.

# February 20

### Ludvig Boltzmann (Theoretical Physics)
### Birthday: 1844, Vienna, Austria

Ludvig Boltzmann studied at the University of Vienna and became professor there of Theoretical Physics. He contributed to the development of the kinetic theory of gases, electromagnetism, and thermodynamics. During his lifetime there was heated scientific debate about his theories and he was depressed by the hostility directed toward his ideas. Ultimately however, Boltzmann's theories were vindicated and his work led to the establishment of statistical mechanics.

# February 20

*Mercury 6 Capsule, Friendship 7 (Space Exploration)*
*Event: 1962, US spacecraft Earth Orbit*

John Glenn orbited Earth on February 20, 1962 - the first American, but not the first person, to do so - in the Project Mercury 6 Space Capsule called Friendship 7. He orbited three times - for a total of 4 hours 55 minutes and splashed down in the Atlantic Ocean. The flight covered about 75,679 miles at a speed of 17,544 miles per hour.

# February 21

*John Mercer (Chemistry)*
*Birthday: 1791, Great Harwood, Lancashire, England*

John Mercer started working at age 9 in the cotton industry. He had no formal education in chemistry, but while working for a calico-printing company he developed the process (using caustic soda on cotton linters) that produces mercerized cotton. He also developed a process that was later used to make artificial silk, and devised photo-chemical processes for printing on fabric.

# February 22

*Heinrich Rudolph Hertz (Physics)*
*Birthday: 1857, Hamburg, Germany*

Heinrich Hertz worked under Helmholtz (August 31) in Berlin and became professor of physics at Karlsruhe. He was a pioneer in radio communication. He established that electric waves behave in the same way as light waves in that they travel in straight lines, and can be reflected, refracted and polarized. In 1895 he produced and detected radio waves over a distance of 60 feet.

# February 23

**Allen McCleod Cormack (Physics)**
**Birthday: 1924, Johannesburg, South Africa**

Alan Cormack studied physics at Cape Town and spent time at the Cavendish (October 10) Laboratory, Cambridge. In the U.S., at Tufts, he published papers, stimulated by work in the Radioisotope Department at the Groot Schurr Hospital in Cape Town, on the mathematical foundation for reassembling X-ray information to construct images of sections of the body in Computerized Axial Tomography.

# February 24

**Thomas Newcomen (Engineering)**
**Christening: 1663, Dartmouth, England**

Thomas Newcomen was a metal worker in Dartmouth and is known as the inventor of the first practical steam engine. In 1712 he built the Newcomen Engine, the first steam operated pump to use a piston and cylinder. This invention was a mainstay of the Industrial Revolution.

*Newcomen Engine*

# February 25

**Phoebus Aaron Theodor Levene (Biochemistry)**
**Birthday: 1869, Sagor, Russia**

Phoebus Levene (originally Levin) studied medicine in St Petersburg and emigrated to the U.S. in 1891. He practiced medicine in New York and, at the same time, studied chemistry at Columbia University. He did pioneer work, based on foundations laid by Emil Fischer (October 9) on nucleic acids, isolated nucleotides, and explored their structure.

# February 26

*Dominique Arago (Physics)*
*Birthday: 1786, Estagel, France*

Dominique Arago is remembered as an astronomer for discovering the solar chromosphere and for accurate measurements of the diameters of the planets, and as a physicist for his work in electromagnetism and light and explanation of the polarization of light in terms of wave theory. Arago entered politics in 1830 and was responsible for ending slavery in the French colonies.

# February 27

*Petrus Artedi (Taxonomy)*
*Birthday: 1705, Angermanland, Sweden*

Artedi's unpublished work *Petri Artedi seuci, medici ichthyologia sive opera omnia de iscibus* was edited by Linnaeus and published posthumously. It established Artedi as the father of fish taxonomy (icthyological taxonomy).

*Alice Hamilton (Medicine)*
*Birthday: 1869, New York City, USA*

Alice Hamilton studied medicine at Michigan and became the first woman member of the Harvard Medical School Faculty. She pioneered research in industrial diseases and studied effects of paint, inks, dyes and explosives. She wrote *Exploring the Dangerous Trades* and her work influenced legislation aimed at correcting industrial working conditions.

# February 28

### Rene Antoine Ferchault de Reaumur (Physics)
*Birthday: 1683, La Rochelle, France*

Rene Reamur was a prodigious talent dubbed by contemporaries as the Pliny of the 18th century. He is generally known today for giving his name to the Reaumur thermometric scale but can also can lay claim to being one of the founders of entomology. His research into technological uses of iron and tinplate steel led to the establishment of steel and tinplate making in France.

### Linus Pauling (Chemistry)
*Birthday: 1901, Portland, Oregon, USA*

Linus Pauling was awarded the 1954 Nobel Prize in Chemistry for his work on the chemical bond and the structure of molecules. In the late 1940s he identified the cause of sickle cell anemia, and in 1961 proposed a molecular model to explain anesthesia. He became popularly renowned for his emphasis on the therapeutic value of Vitamin C in regular, significant, daily doses.

*Sickle cells*

### Peter Brian Medawar (Biology)
*Birthday: 1915, Rio de Janeiro, Brazil*

Peter Medawar's family emigrated to England when he was a young boy. After studying under J.Z. Young (March 18) at Oxford, he made major contributions to immunology working with skin grafts and studying acquired immunological tolerance. He was awarded the 1960 Nobel Prize in Physiology or Medicine. He later studied animal behavior and cancer, and wrote prolifically.

*A scientist must be freely imaginative and yet skeptical, creative and yet a critic...*
*There is poetry in science, but also a lot of bookkeeping."*

*P. B. Medawar (English biologist) February 28*

# February 29

*Charles Pritchard (Astronomy)*
*Birthday: 1808, Alderbury, Shropshire, England*

Charles Pritchard attended St John's College Cambridge and then was headmaster of Clapham Grammar School until 1862. As an astronomer, Pritchard published more than 50 papers, pioneered the use of photographic plates for accurate determination of astronomical positioning, and was responsible for the building of a new observatory in Oxford in 1870.

*Jupiter and Io*

# March

*"The distinctive Western character begins with the Greeks,
who invented the habit of deductive reasoning and the science of geometry."*

*Bertrand Russell (English mathematician) 1872-1970*

*Shasta Daisy*

# Introduction to March

## How Information is Handled is Facinating

I went to a liberal arts college, with a goal of going into foreign work. I decided to study geology, as a means to working abroad, but also studied languages and English. I loved teaching and decided that someday I would probably want to be a university teacher, so I went on to Stanford to get a doctorate in geology. When I completed my doctorate I worked for 11 years in the oil industry in the U.S.: California, New Orleans, and Houston, Texas.

During my stay in New Orleans I read a book called the *Analytical Engine*, telling about computers, their history and the way they were starting to change the world (this was 35 years ago). I realized that Information Handling was what I really, really wanted to do more than geology. Understanding information is fascinating, requiring knowledge about knowledge, and I decided to change my focus to that field. To do so, I set out to learn how to handle oil well data systems through computers. I got deeply into that work, and spent my last year in the industry setting up a research department on the use of computers in geologic exploration.

My work in computers led to an offer to come to UC Davis and explore the applications of computers to medicine. I spent my first ten years working in medical education and human physiology, using computers and other technology to support learning and research in those fields. Since then I have been extending my interest in the characteristics of knowledge, with an emphasis on computer support for distance learning. I have never regretted this change, nor do I regret any of the many areas I have been privileged to study. I love my work and do not look forward at all to retirement.

*Richard Walters*
*Dept. of Computer Science, University of California, Davis*
*May 1997*

# March

| | | | |
|---|---|---|---|
| 1 | Archer John Porter Martin, 1910 - | 16 | Caroline Herschel, 1750 - 1848 |
| 2 | Edward Uhler Condon, 1902 - 1974 | | Georg Simon Ohm, 1789 - 1854 |
| 3 | Alexander Graham Bell, 1847 - 1922 | 17 | William Withering, 1741 - 1799 |
| | Arthur Kornberg, 1918 - | 18 | John Zachary Young, 1907 - 1997 |
| 4 | William Boyd, 1903 - 1983 | | Voskshod 2, 1965 |
| 5 | Gerardus Mercator, 1512 - 1594 | 19 | Walter N. Haworth, 1883 - 1950 |
| 6 | Joseph von Fraunhofer, 1787 - 1826 | 20 | Torbern Olaf Bergman, 1735 - 1784 |
| 7 | Luther Burbank, 1849 - 1926 | 21 | Walter Gilbert, 1932 - |
| 8 | Otto Hahn, 1879 - 1968 | 22 | Robert Andrews Millikan, 1868 - 1953 |
| 9 | Howard Hathaway Aiken, 1900 - 1973 | | Burton Richter, 1931 - |
| 10 | Marcello Malpighi, 1628 - 1694 | 23 | Amelie Emmy Noether, 1882 - 1935 |
| 11 | Urbain Jean Joseph Le Verrier, 1811 - 1877 | | Joanne Simpson, 1923 - |
| | Charlotte Friend, 1921 - 1987 | 24 | John Cowdery Kendrew, 1917 - 1997 |
| 12 | Simon Newcomb, 1835 - 1909 | 25 | Norman Borlaug, 1914 - |
| 13 | Joseph Priestly, 1733 - 1804 | 26 | Bernard Katz, 1911 |
| 14 | Paul Ehrlich, 1854 - 1915 | 27 | Wilhelm Conrad Roentgen, 1845 - 1923 |
| | Albert Einstein, 1879 - 1955 | 28 | Wilhelm Friedrich Kuhne, 1837 - 1900 |
| 15 | Emil Adolph von Behring, 1854 - 1917 | 29 | Edwin Laurentine Drake, 1819 - 1880 |
| | Grace Chisholm Young, 1868 - 1944 | 30 | Bernhard Voldemar Schmidt, 1879 - 1935 |
| | | 31 | Robert Wilhelm Bunsen, 1811 - 1899 |
| | | | Sin-Itiro Tomonaga, 1906 - 1979 |

# March 1

*Archer John Porter Martin (Biochemistry)*
*Birthday: 1910, London, England*

Archer John Porter Martin worked at the Wool Industries Research Associates in Leeds and developed partition chromatography. In 1943 he developed paper chromatography. He was awarded (with R.L.M. Synge) the 1952 Nobel Prize for Chemistry.

# March 2

*Edward Uhler Condon (Physics)*
*Birthday: 1902, Alamagordo, New Mexico, USA*

Edward Condon was distinguished for his research in atomic spectroscopy. With James Franck he developed the Franck-Condon principle. In 1929 he published the book *Quantum Mechanics*. He was associate director, with Oppenheimer (April 22), of the Manhattan Project, and later (1945-51) was director of a U.S.A. Air Force study of unidentified flying objects.

# March 3

*Alexander Graham Bell (Invention)*
*Birthday: 1847, Edinburgh, Scotland*

Alexander Graham Bell started his career as a teacher and then studied medicine. He made a particular study of sound waves and became very interested the mechanics of speech and later in telegraphic communication. He set up a Boston school for training the deaf; founded the professional journal Science; and in 1876 received a patent for his invention of a telephone. This invention arose from his experiments in collaboration with Thomas Watson using electromagnetic principles.

*Alexander Graham Bell*

# March 3

*Arthur Kornberg (Biochemistry)*
*Birthday: 1918, Brooklyn, New York, USA*

Arthur Kornberg was a physician and a biochemist who was educated at the City College in New York and at Rochester. He investigated enzymes and their metabolic activity and studied nucleotide formation in living cells. His work clarified the way in which DNA molecules are built up and replicated. He shared the 1959 Nobel Prize for Physiology or Medicine.

# March 4

*William Boyd (Biochemistry)*
*Birthday: 1903, Dearborn, Montana, USA*

William Boyd was educated at Harvard and studied blood groups at several universities. In 1945 he studied the idea that some plant proteins have blood group specificity. He ground lima beans with sand and obtained an extract with saline. Testing this with samples of human blood cells he found it to have an affinity for antigen A. This became a basis for blood group identification.

# March 5

*Gerardus Mercator (Cartography)*
*Birthday: 1512, Rupelmonde, Flanders*

Gerardus Mercator was a mathematical instrument maker and geographer known primarily for his skill as a map maker and for the 1569 chart of the world using what is now called Mercator's projection. This is a cylindrical projection which increasingly exaggerates east-west distances as one travels north and south from the equator. This is still the most familiar world map projection.

# March 6

*Joseph von Fraunhofer (Optics)*
*Birthday: 1787, Straubing, Bavaria, Germany*

Joseph von Fraunhofer was a partner in the Mechanical-Optical Institute in Germany and studied the refractive properties of glass. He investigated the dark lines of the solar spectrum and proved them to be an intrinsic feature of the spectrum. His work transformed the spectroscope from a scientific toy into an important, precision research instrument.

# March 7

*Luther Burbank (Horticulture)*
*Birthday: 1849, Lancaster, Massachusetts, USA*

Luther Burbank settled, in 1875, in an ideal spot in Santa Rosa, California, and developed a world famous garden. He used his practical experience and Darwinian (February 12) evolutionary theory to crossbreed, hybridize and graft to develop new fruit and vegetable varieties. He worked with prunes, blackberries, gladioli, pears and peaches. He is perhaps most famous for the Shasta daisy, and a spineless cactus which was a valuable cattle food.

*Shasta Daisy*

# March 8

*Otto Hahn (Chemistry)*
*Birthday: 1879, Frankfurt-am-Main, Germany*

Otto Hahn prepared at first for a career as an architect, but then became intensely interested in chemistry. He studied radiochemistry in London in 1904. Returning in 1906 to Germany, he was joined in his research by Lise Meitner (November 7). His work on radioactivity during the next 30 years led to the realization that nuclear fission was possible. He won the 1944 Nobel Prize for Chemistry.

# March 9

*Howard Hathaway Aiken (Engineering)*
*Birthday: 1900, Hoboken, New Jersey, USA*

Howard Aiken studied engineering at the University of Wisconsin and received his doctorate from Harvard. He began his research into computer technology in the 1930s when the field was in its infancy. Aiken became world famous as a result of his work for the U.S. Navy in the development of the first Automatic Sequence Controlled Calculator – the Harvard Mark I – a mechanical device weighing more than 30 tonnes. The Mark II, completed in 1947, and his second ground-breaking achievement, was a fully electronic – but still huge – computing machine.

# March 10

*Marcello Malpighi (Medicine)*
*Baptism: 1628, Bologna, Italy*

Marcello Malpighi graduated in philosophy and medicine. His greatest experiments were done in 1660-61 when he pioneered the study of the structure of the human lung and discovered blood capillaries. He gave his name to part of the kidney (Malpighian bodies). He was also a pioneer in embryology, histology and comparative anatomy.

# March 11

*Urbain Jean Joseph Le Verrier (Astronomy)*
*Birthday: 1811, St Lo, Normandy, France*

Urbain Jean Joseph Le Verrier studied the planet Uranus and postulated in 1846 that the irregularities of its behavior might be due to the presence of an undiscovered planet. He wrote to J.G. Galle at the Berlin Observatory describing the possible position of this new planet. Galle found the planet that we now know as Neptune, and credit was given to Le Verrier for the discovery.

# March 11

*Charlotte Friend (Microbiology)*
*Birthday: 1921, New York City, USA*

Charlotte Friend received her doctorate from Yale in 1950. Her career in cancer research looked not so much for a cure as for an explanation of the causes of cancer. She was the first to discover a direct link between viruses and cancer. Her presentation supporting this at a cancer research conference in 1956 was met with scorn and ridicule, but eventually the virus that had been the focus of her work for years became known as the Friend virus and was acknowledged as one of the viruses that causes leukemia in mice. Friend also demonstrated the possibility of immunizing mice against leukemia.

# March 12

*Simon Newcomb (Mathematics, Astronomy)*
*Birthday: 1835, Wallace, Nova Scotia, Canada*

Simon Newcomb was first a school-teacher and later a professor of mathematics and astronomy at Johns Hopkins University. He made major contributions to the understanding of the movements of the Moon and the planets. He researched historical records of the Moon's movements since 1675, and improved the tables of positions for the Sun's eight known planets.

# March 13

*Joseph Priestly (Chemistry)*
*Birthday: 1733, Birstal, Yorkshire, England*

Joseph Priestly, encouraged by Benjamin Franklin (January 17), set out in 1765 to write a history of electricity. He experimentally tested the facts that he was reporting and published the work in 1767. He then turned to chemistry and discovered what he called dephlogisticated air, which was also discovered by Scheele (December 9) and was later named oxygen by Lavoisier (August 26). Priestly's 1768 *Essay on Government* provided Thomas Jefferson with ideas for the American Declaration of Independence.

# March 14

*Paul Ehrlich (Medicine)*
*Birthday: 1854, Strehlen, Silesia*

Paul Ehrlich, between 1877-81, identified the different types of human white blood cells and distinguished the different leukemias according to the dominant type of white cell. In 1891 he discovered that injection of poisons induced the production of antibodies. He then searched for and found substances lethal to particular organisms, but harmless to the host. He founded chemotherapy.

*Paul Ehrlich*

*Albert Einstein (Theoretical Physics)*
*Birthday: 1879, Ulm, Wurttemburg, Germany*

Albert Einstein ranks as a theoretical physicist with Copernicus, Galileo and Newton in the significance and influence of his works. He is popularly renowned for the equation $E=mc^2$ and was awarded the 1921 Nobel Prize in Physics for his services to Theoretical Physics and especially for his discovery of the law of the photoelectric effect - work that he did in 1905. His major contribution to science was his Theory of Relativity which he published in 1915.

*"The mere formulation of a problem is often far more essential than its solution, which may be merely a matter of mathematical or experimental skill."*

*Albert Einstein (German/American physicist) March 14*

# March 15

*Emil Adolph von Behring (Medicine, Bacteriology)*
*Birthday: 1854, Hansdorf, West Prussia, Germany*

Emile von Behring was awarded a 1901 Nobel Prize for his work on serum therapy, especially its application against diphtheria. He discovered antitoxins and the successful treatment of diphtheria with a mixture of toxin and antitoxin. This was a forerunner of the modern method for preventing rather than curing the disease.

*Grace Chisholm Young (Mathematics)*
*Birthday: 1868, London, England*

Grace Chisholm studied mathematics at Girton College, Cambridge and obtained, in 1893, the equivalent of a first class degree (women could not earn formal degrees at that time at Cambridge.) She then obtained a Ph.D. magna cum laude in Germany. With her husband William Young she published over 200 mathematical articles and several books. In 1915 she won the Cambridge Gamble Prize with a paper on foundations of the calculus.

# March 16

*Caroline Herschel (Mathematics)*
*Birthday: 1750, Hanover, Germany*

Caroline Herschel, sister of musician/astronomer William Herschel (November 15), lived in Bath with her brother and became first his assistant and later an astronomer in her own right. She helped grind and polish mirrors for the telescopes her brother made and then began making her own independent observations. She drew up a catalog of nebulae and discovered eight comets. She was made an honorary member of the Royal Astronomical Society.

# March 16

*Georg Simon Ohm (Physics)*
*Birthday: 1789, Munich, Bavaria*

Georg Ohm was a physicist and investigated electricity. He developed, in 1826, the physical law which bears his name. Ohm's Law states that a precise relationship exists between electrical current, voltage and resistance: the current in a circuit is directly proportional to the applied voltage and inversely proportional to the circuit resistance. (Current equals voltage over resistance.)

# March 17

*William Withering (Medicine, Botany)*
*Birthday: 1741, Wellington, Shropshire, England*

William Withering is known both for his work in medicine and botany. In medicine he is recognized for his *Classic Account of the Foxglove* (1785) in which he describes the use of digitalis derived from the foxglove plant for the treatment of dropsy and heart disease. In botany he is recognized for his 1776 standard work titled *Botanical Arrangement*. He was interested also in mineralogy and the mineral witherite (barium chloride) is named after him.

*Foxglove*

# March 18

*John Zachary Young (Biology)*
*Birthday: 1907, Fishponds, Bristol, England*

John (J.Z.) Young graduated in zoology at Magdalen College, Oxford. His detailed research on the giant nerve fibers of the squid and octopus helped make neurology a more exact science. Later, as Professor of anatomy at University College, London he published companion volumes: *The Life of Vertebrates* and *The Life of Mammals* in which, because the division of biology into embryology, anatomy, physiology etc. was not acceptable to him, he attempted to give combined accounts to show the continuity of life functions and evolution.

# March 18

*Voskshod 2*
*Event: 1965, Soviet Space Flight*

Voskshod 2 made a successful one day flight on March 18, 1965 carrying Soviet cosmonauts Pavel Belyayev (age 39) and Alexei Leonov (age 31). They were the first men to walk (tethered to the space craft) in space. This mission was also the first in which there was manual control of a Soviet spacecraft and the first Soviet manual re-entry and landing.

# March 19

*Walter N. Haworth (Chemistry)*
*Birthday,: 1883, Chorley, Lancashire, England*

Walter Hayworth of Birmingham University in England was awarded the 1937 Nobel Prize in Chemistry for his studies of carbohydrates and for the elucidation of the structure and the artificial synthesis of vitamin C (which he named ascorbic acid.) He determined the molecular structure of glucose and devised the Haworth formula to describe it three dimensionally.

# March 20

*Torbern Olaf Bergman (Chemistry)*
*Birthday: 1735, Catherineberg, Sweden*

Tobern Olaf Bergman appears to have been a self-taught chemist. (His formal studies were in theology and law.) He produced a new chemical classification of minerals. His most creative work, which remained unfinished when he died, was his experimental study of chemical affinity in which he placed acids and bases in sequence depending upon their affinities for each other.

*Flask*

# March 21

*Walter Gilbert (Molecular Biology)*
*Birthday: 1932, Cambridge, Massachusetts, USA*

Walter Gilbert shared the 1980 Nobel Prize in Chemistry for his 1975 development of a method for determining the sequence of bases in nucleic acids such as DNA. He was instrumental in the initiation of the Human Genome Project in the US. With several other scientists he started the biotechnology company Biogen N.V. and became its CEO in 1981.

---

# March 22

*Robert Andrews Millikan (Physics)*
*Birthday: 1868, Morrison, Illinois, USA*

Robert Millikan graduated from Oberlin College in 1891, joined the University of Chicago in 1896, and concentrated his research on the electron. He showed that the electron's charge was a discrete constant, not a statistical average as had been thought. He experimentally confirmed Einstein's photoelectric equation and won the 1923 Nobel Prize in Physics for his work on the elementary charge of electricity and on the photoelectric effect.

*Burton Richter (Physics)*
*Birthday: 1931, Brooklyn, New York, USA*

Burton Richter was working at the Stanford Linear accelerator when he and his research group discovered a new particle, the J/psi. This particle was discovered independently at the Brookhaven National Laboratory by Samuel Tring. Richter and Tring shared the 1976 Nobel Prize in Physics for pioneering work in the discovery of a heavy elementary particle of a new kind.

# March 23

*Amelie Emmy Noether (Mathematics)*
*Birthday: 1882, Erlangen, Bavaria*

Emmy Noether was a founder of abstract algebra and the New Math. She developed a noncomputational approach to mathematics. She developed Noether's Theorem, which was the foundation of quantum physics. A few weeks after her death in 1935, Albert Einstein, in a letter to the New York Times, described her as the most significant, creative mathematical genius.

*Joanne Simpson (Meteorology)*
*Birthday: 1923, Boston, Massachusetts, USA*

Joanne Simpson received her doctorate from the University of Chicago in 1949. Her interests and research areas have been in studying trade winds, hurricanes, and cloud seeding. In the late 1960s and early 1970s Simpson did research on the rain-producing cumulus clouds and developed and tested a mathematical computer model of the characteristics of these clouds.

# March 24

*John Cowdery Kendrew (Physics, Biochemistry)*
*Birthday: 1917, Oxford, England*

John Cowdery Kendrew was a molecular biologist. He used X-ray crystallography to formulate the first three dimensional description of the protein myoglobin. For this achievement he shared (with Max Perutz in Cambridge) the 1962 Nobel Prize in Chemistry for studies of the structure of globular proteins.. He founded the Journal of Molecular Biology in 1959 and published *The Thread of Life* in 1966.

## March 25

*Norman Borlaug (Agriculture)*
*Birthday: 1914, Cresco, Iowa, USA*

Norman Borlaug was an agricultural scientist who developed high yield grains and created a system of plant breeding and crop management now called the Green Revolution. He was awarded the 1970 Nobel Peace Prize for his service to humanity through his agricultural innovations which were credited with helping to alleviate world hunger.

## March 26

*Bernard Katz (Physiology)*
*Birthday: 1911, Leipzig, Germany*

Bernard Katz studied nerve transmission within and between neurons especially those involved in muscle stimulation. He discovered that tiny packets of neurotransmitter (such as acetylcholine) molecules are responsible for much neural activity. These packets are now called vesicles. Katz shared the 1970 Nobel Prize for Physiology or Medicine for discoveries concerning the humoral transmitters in the nerve terminals and the mechanism for their storage, release and inactivation.

## March 27

*Wilhelm Conrad Roentgen (Physics)*
*Birthday: 1845, Lennep, Germany*

William Conrad Roentgen is most famous for his 1895 accidental discovery of X-rays. This discovery had a revolutionary effect on both physics and medicine and Roentgen received the 1901 Nobel Prize in Physics for the extraordinary services he rendered by discovery. Roentgen's real research area was on the specific heat of gases, the heat conductivity of crystals, pyroelectricity, and piezoelectricity.

*Early X-ray*

# March 28

## Wilhelm Friedrich Kuhne (Physiology)
### Birthday: 1837, Hamburg, Germany

William Friedrich Kuhne studied medicine at Berlin. He was a physiologist principally interested in proteins and studied egg albumen coagulation at different temperatures. He coined the term enzyme which in Greek means *in yeast*. He also gave the name visual purple to the photosensitive pigment in the retina of the frog.

# March 29

## Edwin Laurentine Drake (Oil Drilling)
### Birthday: 1819, Bethlehem, Pennsylvania, USA

Edwin Laurentine Drake was a pioneer of oil well drilling. He developed the use of a pipe down to the bedrock to protect the drill-hole of an oil well. He neglected to patent the idea however which was, financially speaking, an unfortunate oversight. At Oil Creek, near Titusville, one of his wells struck oil at 69 feet. This event in August 1859 marked the beginning of the modern petroleum industry.

# March 30

## Bernhard Voldemar Schmidt (Engineering)
### Birthday: 1879, Insel Nargen, Estonia

Bernhard Voldemar Schmidt taught himself lens-grinding and made optical instruments. Later he worked at manufacturing lenses and mirrors for telescopes and cameras. His claim to fame is his astronomical camera that is practically free of chromatic aberration. The largest Schmidt telescope yet built is at Mount Palomar.

# March 31

## Robert Wilhelm Bunsen (Chemistry)
### Birthday: 1811, Gottingen, Germany

Robert Bunsen studied chemistry at Gottingen and became a great experimental chemist pioneering chemical spectroscopy. His work helped Edward Frankland's (January 18) development of the concept of valency. He detected and isolated two new metals, rubidium and cesium. He wrote a book on gas analysis and made several inventions but probably played only a small part in the invention of the Bunsen Burner.

## Sin-Itiro Tomonaga (Physics)
### Birthday: 1906, Tokyo, Japan

Sin-Itiro Tomonaga was a pioneer in the field of quantum electrodynamics. His theory concerning subatomic particles, which was consistent with the theory of relativity, was formulated at the same time as, independently, Richard Feynman (May 11) and Julian Schwinger reached the same conclusions. The three shared the 1965 Nobel Prize in Physics for fundamental work in quantum electrodynamics, with deep-ploughing consequences for the physics of elementary particles.

*Cubic closest packing of Rubidium*

# April

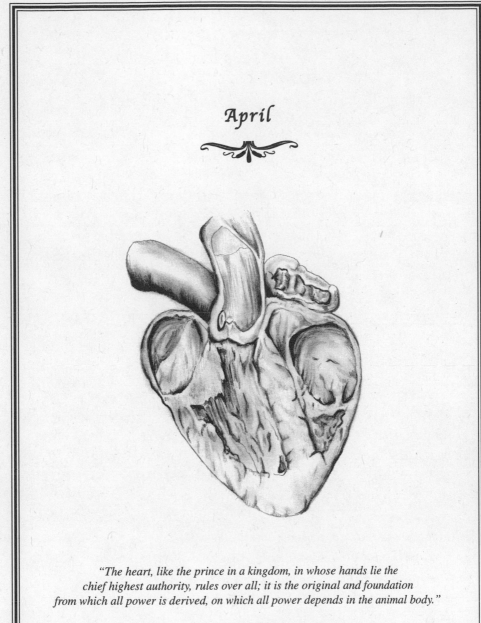

*"The heart, like the prince in a kingdom, in whose hands lie the chief highest authority, rules over all; it is the original and foundation from which all power is derived, on which all power depends in the animal body."*

*William Harvey (English physician) April 1*

*Human heart*

# Introduction to April

## *Appreciating Nature Through Science*

I have spent much of my scientific career studying and writing about gorilla behavior. Anyone who sees gorillas in the wild is awed by them. Words such as 'impressive' 'magnificent', and of course, 'gentle giant' are used to describe the animals. But many people think that being a scientist somehow neutralizes or sterilizes this experience; that closely scrutinizing gorillas and then reducing them to columns of numbers and points on graphs, as I have done in my work, takes away their mystery. This is like saying that to know about the fish in the oceans detracts from the beauty of the sun setting on the water. Exploring the depths does not diminish the reflections from the surface. It simply adds to the experience.

I never ceased to be moved by wild gorillas, even though I spent months on end in their presence, detailing their behavior. But science allowed me to be more than a wonder-struck observer. When I saw a 300 pound adult male, for example, I also saw the force of competition for mates that drives males to grow bigger than females. I could ponder the relation between this competition and the animals' social system or their ecology. I could understand how and why they differed from other monkeys and apes, such as chimpanzees. And I could begin to see how and why they differed from, or were similar to, ourselves. That's another good reason to study the lives of other creatures, apart from the pure fascination of it. It shows us where we fit into the rest of nature. It keeps us humble.

*Kelly Stewart*
*Dept. of Anthropology, University of California, Davis*
*April 1997*

# April

| | | | |
|---|---|---|---|
| 1 | William Harvey, 1578 - 1657 | 16 | Wilbur Wright, 1867 - 1912 |
| | Sophie Germain, 1776 - 1831 | | Marie Maynard Daly, 1921 - |
| 2 | Francesco Maria Grimaldi, 1618 - 1663 | 17 | Ernest Henry Starling, 1866 - 1927 |
| 3 | Jane Goodall, 1934 - | 18 | Eugene Jules Houdry, 1892 - 1962 |
| 4 | Charles W. Siemens, 1823 - 1883 | 19 | Glen Theodore Seaborg, 1912 - |
| 5 | Joseph Lister, 1827 - 1912 | 20 | Marc Seguin, 1786 - 1875 |
| | Hattie Alexander, 1901 - 1968 | 21 | John Muir, 1838 - 1914 |
| 6 | James Dewey Watson, 1928 - | 22 | Robert Oppenheimer, 1904 - 1967 |
| 7 | Eric Ivar Fredholm, 1866 - 1927 | | Rita Levi-Montalcini, 1909 - |
| 8 | Harvey Cushing, 1869 - 1939 | 23 | Max Karl E. Ludvig Planck, 1858 - 1947 |
| 9 | John Presper Eckert, 1919 - | | Johannes A. Grib Fibiger, 1867 - 1928 |
| 10 | Robert Burns Woodward, 1917 - 1979 | 24 | Edmund Cartwright, 1743 - 1823 |
| 11 | Donald Howard Menzel, 1901 - 1976 | 25 | Guglielmo Marconi, 1874 - 1937 |
| 12 | Carl L. F. Lindemann, 1852 - 1939 | 26 | John James Audubon, 1785 - 1851 |
| | Vostock I, 1961 | 27 | Samuel F. Breese Morse, 1791 - 1872 |
| | Shuttle Columbia, 1981 | 28 | Jan Hendrik Oort, 1900 - 1992 |
| 13 | Richard Trevithick, 1771 - 1833 | | Eugene Merle Shoemaker, 1928 - 1997 |
| 14 | Christiaan Huygens, 1629 - 1695 | 29 | Jules Henri Poincare, 1854 - 1912 |
| | RMS Titanic, 1912 | | Betsy Ancker-Johnson, 1927 - |
| 15 | Leonardo da Vinci, 1452 - 1519 | 30 | Karl Friederick Gauss, 1777 - 1855 |
| | Leonhard Euler, 1707 - 1783 | | |
| | Nicolaas Tinbergen, 1907 - 1988 | | |

# April 1

### William Harvey (Medicine)
### Birthday: 1578, Folkstone, Kent, England

William Harvey is credited with the discovery of the fact that blood circulates around the body and through the heart. In his *Anatomical Treatise on the Movement of the Heart and Blood in Animals* (1628) he described the pumping action of the heart and showed that the right ventricle supplies the pulmonary circulation and the left ventricle the rest of the body.

### Sophie Germain (Mathematics)
### Birthday: 1776, Paris, France

Sophie Germain taught herself mathematics from books in her father's library. While she was not allowed to enroll in the Ecole Polytechnique, she studied the lecture notes for several of the courses, was encouraged by Joseph Lagrange who had organized the mathematics department there, and corresponded with the mathematician K.F. Gauss (April 30). She is best known for her work in number theory and the vibration of elastic surfaces.

# April 2

### Francesco Maria Grimaldi (Physics)
### Birthday: 1618, Bologna, Italy

Francesco Grimaldi discovered diffraction of light. He was a Jesuit scholar and teacher and his early work as an assistant included making observations for a detailed map of the Moon. He performed some fundamental experiments in physics and in his book *Physico-mathesis de Lumine* he describes light as a substantial entity with undulatory properties.

# April 3

### Jane Goodall (Ethology)
### Birthday: 1934, London, England

Jane Goodall is respected for her ground-breaking field studies of the chimpanzees of the Gombe Stream Reserve in Tanzania, Africa. She is credited with the first recorded observations of chimpanzees eating meat and making and using tools. As a result of Goodall's observations, scientists have had to redefine some characteristics once considered uniquely human.

# April 4

### Charles William Siemens (Engineering)
### Birthday: 1823, Lenthe, Hanover, Germany

Charles Siemens was born and educated in Germany, and lived in England after age 20. There he pioneered the open hearth steel furnace which, by 1900, exceeded the Bessemer furnaces in steel production throughout the world. Siemens also designed a cable-laying steamship which laid the Atlantic cable of 1874, and was involved in the building of the Portrush Electric Railway in Ireland.

# April 5

### Joseph Lister (Medicine)
### Birthday: 1827, Upton, Essex, England

Joseph Lister was the founder of antiseptic surgery. General anesthesia was introduced in 1846 but sepsis was an increasing problem. Mortality after amputations was 40 to 60 percent. It was believed that sepsis was caused simply by air, but after much experimentation Lister discovered that carbolic acid destroyed the tiny airborne causative-organisms described by Louis Pasteur (December 27).

# April 5

*Hattie Alexander (Medicine, Microbiology)*
*Birthday: 1901, Baltimore, Maryland, USA*

Hattie Alexander obtained her MD from Johns Hopkins in 1930 and became interested in influenzal meningitis. Her research efforts were directed toward a serum against this infection and she produced a rabbit serum that reduced infant mortality from this type of meningitis by 80 percent. Alexander came to realize that mutation of bacteria allowed resistance to antibiotics.

# April 6

*James Dewey Watson (Molecular Biology)*
*Birthday: 1928, Chicago, Illinois, USA*

James Watson entered the University of Chicago at age 15 and earned a degree in Zoology in 1947. At Indiana, his Ph.D. thesis concerned the effect of X-rays on phage lysis. He then became interested in genetics (in finding out the secret of the gene.) He went on to study the molecular structure of proteins and, in Cambridge, England, discovered the structure of DNA with Francis Crick (June 8) and Maurice Wilkinson, and shared the 1962 Nobel Prize for Physiology or Medicine for discoveries concerning the molecular structure of nuclear acids and its significance for information transfer in living material.

# April 7

*Eric Ivar Fredholm (Mathematics)*
*Birthday: 1866, Stockholm, Sweden*

Eric Ivar Fredholm's particular interest was the solution of problems of practical mechanics. As professor of theoretical physics at Stockholm university, Fredholm founded the modern theory of integral equations. He first published a paper on this subject in 1903. His results were used by David Hilbert (1862-1943) who extended them in developing his own theories on quantum mechanics.

# April 8

*Harvey Cushing (Medicine)*
*Birthday: 1869, Cleveland, Ohio, USA*

Harvey Cushing studied at Yale and Harvard. He became especially interested in intercranial tumors and founded a new discipline of neurosurgery based on history taking and examination. His surgeries lasted for many hours but were increasingly successful, an unusual event in those days. He was also interested in the functions and diseases of the pituitary gland. A disease of the pituitary now bears his name.

# April 9

*John Presper Eckert (Physics)*
*Birthday: 1919, Philadelphia, Pennsylvania, USA*

John Presper Eckert graduated from the Moore School of Electrical Engineering at Pennsylvania where he worked, with John Mauchly, on the first electronic digital computer. The machine used punched cards for data input and filled a 50ft x 30ft room. It was known as ENIAC. In 1948 Eckert and Mauchly formed their own company and developed the BINAC machine which stored data on magnetic tape, and later, UNIVAC. Their work was critically important in the computer field.

# April 10

*Robert Burns Woodward (Organic Chemistry)*
*Birthday: 1917, Boston, Massachusetts, USA*

Robert Burns Woodward spent most of his career at Harvard University. He was awarded the 1965 Nobel prize in Chemistry for the scope of his contributions to chemistry. He explained the structures of many organic compounds including strychnine, terramycin, tetrodotoxin and ferrocene. He synthesized quinine, cholesterol, cortisone, lysergic acid, chlorophyll and vitamin B12.

*Strychnine*

# April 11

*Donald Howard Menzel (Astronomy)*
*Birthday: 1901, Florence, Colorado, USA*

Donald Howard Menzel graduated from Denver and went to Princeton and later Harvard. A physicist, he combined atomic physics with astronomy. His favorite subject was the Sun and he established a new theoretical approach to the structure of its envelope of gases. His work on the spectrum of the solar chromosphere had a strong influence on subsequent solar astronomy.

# April 12

*Carl Louis Ferdinand Lindemann (Mathematics)*
*Birthday: 1852, Hanover, Germany*

Carl Louis Ferdinand Lindemann received his doctorate from Erlangen University and was Professor of Mathematics at Konigsburg for most of his career. He translated and edited the writings of Henri Poincare (April 29) making his mathematics known in Germany. Lindemann's own reputation rests mainly on his discussion and elucidation of $\Pi$ (pi).

*Vostock I*
*Event: 1961, Soviet space flight*

*Vostock I* was piloted by twenty-seven-year-old Russian cosmonaut Yuri Gagarin, the first man in space when, on April 12, 1961 he flew in the spacecraft on man's first orbital flight. The flight lasted for one hour and forty-eight minutes.

*Vostock I*

## April 12

*Shuttle Columbia*
*Event: 1981, U.S. orbital space flight*

The *Shuttle Columbia*, launched for its first orbital flight on April 12, 1981 was piloted by U.S. astronauts John Young and Robert Crippen. The flight lasted two and a half days and landed at Edwards Airforce Base in California.

## April 13

*Richard Trevithick (Engineering)*
*Birthday: 1771, Illogan, Cornwall, England*

Richard Trevithick built a double-acting high-pressure steam engine when James Watt's (January 19) master patent expired in 1800. The increased steam pressure represented an important innovation. He also designed with steam engines a road carriage, a dredger and a threshing machine. His inventions were significant, but he died penniless.

## April 14

*Christiaan Huygens (Mathematics)*
*Birthday: 1629, The Hague, Netherlands*

Christiaan Huygens studied law at Leiden before studying mathematics and science. He discovered the rings of Saturn and, through his major contributions in optics and dynamics, is recognized as one of the great scientists of his time. He developed 'laws of collision' using Galileo's (February 18) 'law of falling bodies' and in his celebrated *Horologium Oscillatorium* (1673) provided explanations concerning the pendulum and centrifugal force. He believed that light was a vibratory motion in the ether and explained the phenomenon of the double refraction in crystals of Iceland Spa.

# April 14

*RMS Titanic*
*Event: 1912, Atlantic Ocean*

RMS *Titanic*, the "unsinkable", trans-atlantic steamer, struck an ice-berg on April 14, 1912 and sank on its maiden voyage from England to New York. The ship, designed by Thomas Andrews, had a steel hull with 16 watertight compartments. Steel production was not yet a perfected art and it is now thought that the steel of the hull might have been brittle due to crystallization.

# April 15

*Leonardo da Vinci (Engineering, Anatomy, Invention)*
*Birthday: 1452, Vinci, near Florence, Italy*

Leonardo da Vinci's scientific investigations are recorded in his notebooks which he wrote in a unique (backwards) hand. He had ideas about many phenomena, invented instruments for measuring wind-force and the speed of ships. Fascinated by the flight of birds he observed this in great detail and designed a machine for human flight as well as other mechanical devices. His anatomical drawings, based on dissection, are beautiful and accurate but remained unknown until centuries after his death.

*Leonhard Euler (Mathematics)*
*Birthday: 1707, Basle, Switzerland*

Leonard Euler is described as the most prolific and greatest all-round mathematician of all time. He was educated at Basle and was friendly with the prestigious Daniel Bernoulli (February 8) family. Like other mathematical geniuses he had a phenomenal memory. His first love was geometry, but his greatest contribution was as the founder of modern analysis. He became blind in 1771 but continued to work and completed his famous treatise on celestial mechanics by dictating the text to his daughter.

# April 15

## Nicolaas Tinbergen (Ethology)
### Birthday: 1907, The Hague, Netherlands

Nicolaas Tinbergen was a zoologist, animal psychologist, and pioneer in the field of ethology. He studied stimulus response processes in wasps, fish and gulls. He shared the 1973 Nobel Prize in Physiology or Medicine (with Karl Frisch and Konrad Lorenz) for discoveries concerning organization and elicitation of individual and social behavior patterns in animals.

*Wright Brothers*

# April 16

## Wilbur Wright (Invention)
### Birthday: 1867, Millville, Indiana, USA

Wilbur Wright, with his brother Orville (August 19), designed and constructed the first successful airplane. In 1900 and 1901 the brothers tested their first gliders. By 1905 they had experimented with planes powered by a petrol engine and in October 1905 they flew a powered aircraft on a 24-mile circuit. They patented their invention in 1906.

## Marie Maynard Daly (Biochemistry)
### Birthday: 1921, Corona, NY, USA

The first black female to earn a Ph.D. in Chemistry, Marie Maynard Daly received her degree from Columbia University in 1948. She served as an Investigator for the American Heart Association (1958-63) and Cancer Scientist for the Health Research Council of New York (1962-72). Daly's area of research focused on nucleic acids.

*"We were on the point of abandoning our work when the book of Mouillard fell into our hands, and we continued with the results you know."*

*Wilbur Wright (American aviation inventor) April 16*

# April 17

*Ernest Henry Starling (Physiology)*
*Birthday: 1866, London, England*

Ernest Henry Starling is known for his discovery, with co-worker W.M. Bayliss (May 2), of secretin (secreted by the small intestine) which causes the release of pancreatic juices. He coined the name hormone for any chemical that is produced in one part of the body and causes a specific effect in another. Starling also studied the circulatory and lymphatic systems.

# April 18

*Eugene Jules Houdry (Chemical Engineering)*
*Birthday: 1892, Domont, Paris, France*

Eugene Jules Houdry trained in Paris as a mechanical engineer but became interested in attempts to produce petrol from lignite by catalytic hydrogenation. In 1936-7 after emigrating to America, Houdry used a hydrated aluminum silicate catalyst to produce high octane petrol. His new process made 100 octane aviation fuel available to the Allies during World War 2.

# April 19

*Glen Theodore Seaborg (Chemistry)*
*Birthday: 1912, Ishpeming, Michigan, USA*

Glen Seaborg and his colleagues identified, in 1940, isotopes of plutonium and demonstrated the nuclear fission of plutonium-239 when bombarded with low energy neutrons. Between 1940 and 1974 Seaborg was involved in the identification and isolation of nine transuranic elements. He shared the 1951 Nobel Prize in Chemistry for discoveries in the chemistry of the transuranium elements.

# April 20

*Marc Seguin (Civil Engineering)*
*Birthday: 1786, Ofen, Hungary*

Marc Seguin, a self-taught designer/builder, constructed suspension bridges and designed and built at Geneva, Switzerland, the first such bridge in Europe. His other great achievement was the invention of a multitubular boiler for the engines of the Lyons and St Etienne Railway in France.

# April 21

*John Muir (Naturalist)*
*Birthday: 1838, Dunbar, Scotland*

John Muir attended Wisconsin University but took no degree because he refused to follow a curriculum of courses. He was avidly interested in nature, made extensive journeys on foot, and kept voluminous daily journals. Yosemite Valley in California was a focus of interest for 11 years. He wrote articles and books and advocated for the protection of public lands.

*Half Dome in Yosemite*

# April 22

*Robert Oppenheimer (Physics)*
*Birthday: 1904, New York City, USA*

Robert Oppenheimer was a brilliant theoretical physicist and a powerful problem solver with a gift for verbal communication. He worked on the quantum theory of molecules and correctly postulated the existence of the positron before it had been discovered. During World War II, in 1942, he was asked to set up and lead a team working to develop an atomic bomb before the Germans did so. He set up his team at Los Alamos, New Mexico, and they successfully developed and tested the bomb.

# April 22

### Rita Levi-Montalcini (Embryology)
*Birthday: 1909, Turin, Italy*

Rita Levi-Montalcini shared the 1986 Nobel Prize in Physiology or Medicine for the discovery of the stimulating substance, that she called the nerve growth factor (NGF), involved in the growth of nerve cells. This factor may play a role in such central nervous system diseases as Alzheimer's.

*Rita Levi-Montalcini*

---

# April 23

### Max Karl Ernst Ludvig Planck (Physics)
*Birthday: 1858, Kiel, Germany*

Max Planck devoted his professional career to the study of physics although he had a passion for music and was a fine pianist. He investigated dissociation of gases, osmotic pressure, and the lowering of freezing point in solutions. He was intrigued by blackbody radiation and, with a flash of intuition, formulated an equation that explained the phenomenon in terms of light that was not a continuous wave but tiny parcels of energy that he called quanta. He was awarded a 1918 Nobel Prize in Physics for the discovery of energy quanta.

### Johannes Andreas Grib Fibiger (Medicine)
*Birthday: 1867, Silkeborg, Denmark*

Johannes Andreas Grib Fibiger was qualified in medicine and did research into causes of cancer. Noting that a nematode worm was associated with tumerous growths in the stomachs of three rats, he successfully induced cancerous growths by feeding rats the larval stage of the worm. By providing a way to induce cancer artificially, Fibiger moved the study of cancer by a giant leap. He was awarded a 1926 Nobel Prize in Physiology or Medicine for his discovery of the Spiroptera carcinoma.

# April 24

*Edmund Cartwright (Invention)*
*Birthday: 1743, Marnham, Nottingham, England*

Edmund Cartwright was educated at University College Oxford, and elected a Fellow of Magdalen. Cartwright was an inventor. He first invented a mechanical loom from which modern looms have evolved. His other inventions included a wool-combing machine (1789), bread making and brick making machines, and a machine to make ropes (1792). In 1793 he designed a reaping machine.

# April 25

*Guglielmo Marconi (Invention)*
*Birthday: 1874, Bologna, Italy*

Guglielmo Marconi, educated by private tutors, had inventive talent and a flair for recognizing practical applications in the inventions of others. In 1895 he built a radio with a range of more than a mile. In 1897 with a patent in England for wireless telegraphy, he formed a Wireless Telegraphy office in London. His first transatlantic signals were made in December 1901. He shared the 1909 Nobel Prize in Physics in recognition of contributions to the development of wireless telegraphy.

# April 26

*John James Audubon (Natural History, Art)*
*Birthday: 1785, Les Cayes, Santo Domingo*

John James Audubon was brought up in France and very early developed a passion for natural history and drawing. He emigrated to America and traveled widely collecting and painting birds and other animals. No American publisher would take his work so he went to England and arranged for the Havells to engrave his plates for publication. Later works were published in the U.S..

# April 27

*Samuel Finley Breese Morse (Invention)*
*Birthday: 1791, Charlestown, Massachusetts, USA*

Samuel Morse studied in London, France and Italy and, until he was forty, his life was devoted to painting. He did however invent a piston pump and a machine for cutting marble. Then, in 1832, he designed a transmitting device based on interruption of an electric circuit, a receiving apparatus with an electro magnet and pen, and the dot-and-dash code that bears his name today.

---

# April 28

*Oort Cloud around*
*our Solar System*

*Jan Hendrik Oort (Astronomy)*
*Birthday: 1900, Franeker, Netherlands*

Jan Hendrik Oort spent his working life as an astronomer at Leiden Observatory in the Netherlands. He is regarded as the Netherlands greatest astronomer. In 1927 he showed that our galaxy is rotating. He and his colleagues mapped the spiral arms of the galaxy and studied its previously unseen center. He postulated a cloud of comets surrounding our solar system (the Oort Cloud.)

*Eugene Merle Shoemaker (Planetary Geology)*
*Birthday: 1928, Los Angeles, California, USA*

Eugene Merle Shoemaker was s chief founder of planetary geology working closely with his wife Carolyn Spellman. They identified 32 comets including Comet Shoemaker-Levy 9 which dazzled sky gazers in July 1994 as it crashed into Jupiter. In the late 1950s Shoemaker presented the world with his theory that a huge meteor crater near Winslow, Arizona was formed by a meteorite impact more than 50,000 years ago. His theory added weight to the possibility that the extinction of the dinosaurs was caused by a similar impact.

# April 29

*Jules Henri Poincare (Mathematics)*
*Birthday: 1854, Nancy, France*

Jules Henri Poincare qualified as a mining engineer but his thesis was mathematical. In his contributions to science as a mathematician he developed ideas on both mathematical techniques and pure mathematics. He made contributions with papers on light, electricity, capillarity, thermodynamics, heat, elasticity and telegraphy, and to the development of relativity theory. He believed that nothing could exceed the speed of light and that it would never be possible to demonstrate anything but relative velocities because absolute motion is undetectable.

*Betsy Ancker-Johnson (Physics, Engineering)*
*Birthday: 1927, St. Louis, Missouri, USA*

Betsy Ancker-Johnson studied physics at Wellesley College, and earned her doctorate from Tubingen University in Germany in 1953. She made important contributions in the field of solid state physics and holds several patents. Her work has implications for computer technology and the extraction of aluminum and other elements from their ores. Ancker-Johnson was appointed Assistant Secretary of Commerce between 1973 and 1977.

# April 30

*Karl Friederick Gauss (Mathematics, Astronomy, Physics)*
*Birthday: 1777, Brunswick, Germany*

Karl Friederick Gauss was unusual in his extraordinary facility for complex mental computation. He made many ground breaking contributions in mathematics and made possible much of the mathematical physics of the next century. His work has been compared in stature to Archimedes and Newton (December 25) but is said to have had a greater range. He developed the 'law of quadratic reciprocity' and the 'method of least squares'. His new concept of congruence enabled him to simplify earlier proofs and theorems.

*"Science is facts. Just as houses are made of stones, so is science made of facts.*
*But a pile of stones is not a house and a collection of facts is not necessarily science."*

*Jules Henrie Poincaré (French mathematician) April 29*

# May

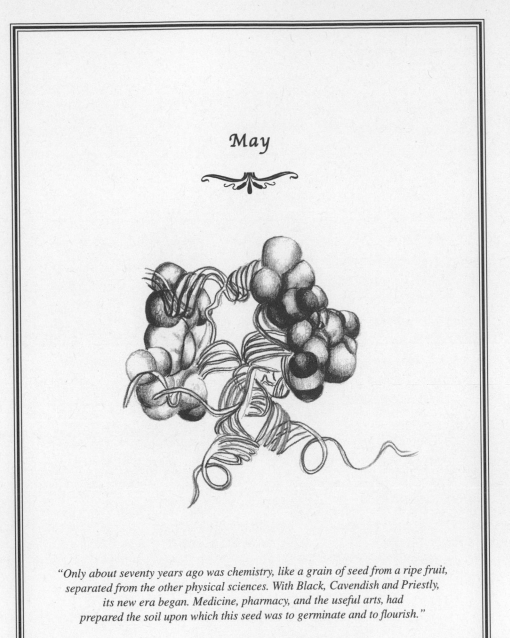

*"Only about seventy years ago was chemistry, like a grain of seed from a ripe fruit, separated from the other physical sciences. With Black, Cavendish and Priestly, its new era began. Medicine, pharmacy, and the useful arts, had prepared the soil upon which this seed was to germinate and to flourish."*

Justus von Liebig (German chemist) May 12

*Hemoglobin*

# Introduction to May

## Real Science

Science as practiced in the lab, in the field or at the computer is very different than science as presented in a textbook. Textbook science is neat, orderly, logical, systematic and coherent. It comes in a nice package, wrapped in colorful paper and tied with a pretty ribbon.

Real science is messy, incomplete, illogical and disorganized. More often than not, the data points do not all fall on a straight line, certain key experiments may not yet have been done, slightly different studies give very different results, critically important parameters can't be measured, and although most scientists are reluctant to admit it, their interpretations may be influenced by their preconceptions and their biases.

Somehow, out of all of this chaos, there emerges a consensus that becomes the temporarily prevailing paradigm. At this point, real science is transformed into textbook science. What is lost in the process is the sense of excitement and challenge that scientists experience when they are at the frontiers of discovery.

If we are to develop a deeper public appreciation and understanding of real science, we need to provide students with a more accurate view of how science is done and why scientists enjoy doing it.

*Kenneth L. Verosub*
*Dept of Geology, University of California, Davis*
*June 1997*

# May

| | | | |
|---|---|---|---|
| 1 | Alexander Williamson, 1824 - 1904 | 15 | Pierre Curie, 1859 - 1906 |
| | Elizabeth Shull Russell, 1913 - | | Dorothy Andersen, 1901 - 1963 |
| | Evelyn Boyd Granville, 1924 | 16 | Maria Gaetana Agnesi, 1718 - 1799 |
| 2 | William Maddock Bayliss, 1860 - 1924 | | Ilya (Elie) Me(t)chnikov, 1845 - 1916 |
| 3 | John Scott Haldane, 1860 - 1936 | 17 | Edward Jenner, 1749 - 1823 |
| 4 | Thomas Henry Huxley, 1825 - 1895 | 18 | Thomas Midgley, 1889 - 1944 |
| 5 | John William Draper, 1811 - 1882 | 19 | Max Ferdinand Perutz, 1914 - |
| 6 | Sigmund Freud, 1856 - 1939 | 20 | Eduard Buchner, 1860 - 1917 |
| | Ann Haven Morgan, 1882 1966 | 21 | Lyon Playfair, 1818 - 1898 |
| 7 | Edwin Land, 1909 - 1991 | 22 | William Sturgeon, 1783 - 1850 |
| 8 | Nevil Vincent Sidgwick, 1873 - 1952 | 23 | Carl Linnaeus, 1707 - 1778 |
| 9 | Carl Gustaf Patrik De Laval, 1845 - 1913 | | John Bardeen, 1908 - 1991 |
| 10 | Gaspard Monge, 1746 - 1818 | 24 | William Gilbert, 1544 - 1603 |
| 11 | Richard Feynman, 1918 - 1988 | 25 | Pieter Zeeman, 1865 - 1943 |
| 12 | Justus Von Liebig, 1803 - 1873 | 26 | Sally Kristin Ride, 1951 - |
| | Florence Nightingale, 1820 - 1910 | 27 | Rachel Louise Carson, 1907 - 1964 |
| | Dorothy Crowfoot Hodgkin, 1910 - 1994 | 28 | Jean Louis R. Agassiz, 1807 - 1873 |
| 13 | Ronald Ross, 1857 - 1932 | 29 | Chien-Shiun Wu, 1912 - 1997 |
| 14 | Gabriel Fahrenheit, 1686 - 1736 | | Paul Ralph Ehrlich, 1932 - |
| | Charlotte Auerbach, 1899 - 1994 | 30 | Julius Axelrod, 1912 - |
| | | 31 | Cagniard de Latour, 1777 - 1859 |

# May 1

*Alexander Williamson (Chemistry)*
*Birthday: 1824, Wandsworth, London, England*

Alexander Williamson studied chemistry at Heidelberg and had a private lab in Paris. Then, in 1849, he was made a professor of analytic chemistry at University College London. He was known as an inspiring teacher but is best remembered for his synthesis of ethers which clarified the relationship between ethers and alcohols.

*Elizabeth Shull Russell (Genetics)*
*Birthday: 1913, Ann Arbor, Michigan, USA*

Elizabeth Russell earned her doctorate at the University of Chicago. She worked at the Roscoe B. Jackson Laboratory in Bar Harbor, Maine from 1935 and after the fire that destroyed the facility in 1947 was responsible for rebuilding the collection of special genetic strains represented amongst the 90,000 mice that had been destroyed when the building burned down. As is the case with so many scientists, it is the sum total of her work adding to the store of scientific knowledge that made Russel a scientist of stature although she did make important discoveries in the field of genetics.

*Evelyn Boyd Granville (Mathematics)*
*Birthday: 1924, Washington, D.C. USA*

Evelyn Boyd Granville received a Ph.D. in Mathematics from Yale in 1949. She was one of the first two Black women to receive doctorates in Mathematics in the United States. Her dissertation was *On Laguerre Series in the Complex Domain.*

*"The great tragedy of science—*
*the slaying of a beautiful hypothesis by an ugly fact."*

*Thomas Henry Huxley (English biologist) May 4*

## May 2

### William Maddock Bayliss (Physiology)
### Birthday: 1860, Wolverhampton, Staffordshire, England

William Maddock Bayliss, with E.H. Starling (April 17), discovered secretin in 1902. Also in collaboration with Starling, he studied venous and arterial pressures. Independently, Bayliss studied vaso-motor reflexes and the psycho-chemical aspects of physiology. During World War 1 he used saline injections for the amelioration of surgical shock.

## May 3

### John Scott Haldane (Physiology)
### Birthday: 1860, Edinburgh, Scotland

John Scott Haldane was an authority on respiration, especially under special conditions such as down a coal mine or on a mountain. He studied black damp in mines and choke damp in wells. In 1911 he led an expedition to Pikes Peak in Colorado, U.S. to study acclimatization and cyanosis at high altitudes.

*Coal Miner*

## May 4

### Thomas Henry Huxley (Zoology)
### Birthday: 1825, Ealing, London, England

Thomas Henry Huxley produced more than 150 research papers in a vast array of subjects although the majority were zoological or paleontological. He wrote ten scientific textbooks. Huxley was a strong supporter of Charles Darwin (February 12) and, being eloquent and well respected, had influence in breaking down the religious opposition to Darwin's evolution theory.

# May 5

*John William Draper (Chemistry)*
*Birthday: 1811, St Helens, Lancashire, England*

John William Draper studied chemistry at London University and then emigrated to America. He founded the New York University School of Medicine and became its president. In 1839 he became very interested in photography. In that year he took one of the first photo portraits in sunlight (a 65 second exposure). Shortly after, he took the first known photograph of the moon.

# May 6

*Sigmund Freud (Psychology)*
*Birthday: 1856, Freiberg, Moravia*

Sigmund Freud was the founder/creator of professional psychoanalysis. He trained in medicine and did neurological research. Then he embarked on clinical work on nervous disorders and created a psychoanalytic method based on free association and dream analysis. He distinguished between two levels of mental functioning, a primitive, subconscious level, and a logical, thinking level.

*Ann Haven Morgan (Zoology)*
*Birthday: 1882, Waterford, Connecticut, USA*

Ann Haven Morgan studied at Wellesley and Cornell and spent her professional life as a professor at Mt. Holyoke College. Her interests focused on aquatic insects, hibernation, conservation and ecology. Her *Field Book of Ponds and Streams* (1930) is an important work for both amateurs and professionals in the field.

# May 7

*Edwin Land (Chemistry)*
*Birthday: 1909, Bridgeport, Connecticut, USA*

Edwin Land was the inventor of the Polaroid process of instant photography. His first interest and commercial venture was the manufacture of polarizing materials (for sunglasses and gun sights). An instant picture process had already been developed by Agfa and Gevaert but Land's Polaroid process, with better quality pictures, went on sale with special cameras in 1948.

*Polaroid Camera*

# May 8

*Nevil Vincent Sidgwick (Chemistry)*
*Birthday: 1873, Oxford, England*

Nevil Vincent Sidgwick was greatly influenced by Ernest Rutherford (August 30) whom he met in 1914 and who stimulated his interest in atomic structure. He began to interpret chemical reactions in terms of the new electron theories being developed in physics. In 1927 his work *The Electronic Theory of Valency* established his reputation as a major player in the development of the valency theory introduced by Frankland (January 18).

# May 9

*Carl Gustaf Patrik De Laval (Invention)*
*Birthday: 1845, Orsa, Sweden*

Carl Gustav Patrik De Laval was educated at the Stockholm Technical Institute and Uppsala University. In 1878 he invented a high-speed centrifugal cream separator which was put into use in dairies worldwide. His best achievement (of several thousand inventions) was probably his improvement to the design of the steam turbine.

# May 10

*Gaspard Monge (Mathematics)*
*Birthday: 1746, Beaune, France*

Gaspard Monge was the founder of descriptive geometry. He combined methods of synthetic and analytic geometry, and laid groundwork for methods of modern engineering drawing. He also worked on the application of calculus to curve and surfaces in three dimensions. He was a close friend of Napoleon Bonaparte.

---

# May 11

*Richard Feynman (Physics)*
*Birthday: 1918, New York, NY, USA*

Richard Feynman studied at MIT and Princeton and in 1945 was part of the team at Los Alamos (July 16) working on the atomic bomb. His research has been in quantum electrodynamics and his major contribution was the simplification of the study of interaction among electrons, protons and radiation. He shared the 1965 Nobel Prize in Physics with Schwinger and Tomonaga (March 31).

---

# May 12

*Justus Von Liebig (Chemistry)*
*Birthday: 1803, Darmstadt, Germany*

Justus Von Liebig is considered a great chemist. His early work was in classical organic chemistry but he became interested in the chemistry of living things and in agriculture. His first important work found that silver cyanate and fulminate had the same elementary analysis. But the concept of isomers was not yet known and Liebig's finding was not explained.

# May 12

*Florence Nightingale (Mathematics)*
*Birthday: 1820, Florence, Italy*

Florence Nightingale, an Englishwoman born in Italy, is popularly known as a pioneer of nursing reform. It is less well known that she was accomplished in mathematics. While in Scutari, during the Crimean War, she collected mortality rate and other data as a tool for improving conditions in hospitals, and represented them statistically. She invented polar-area charts. She was made an honorary member of the American Statistical Association in 1874.

*Dorothy Crowfoot Hodgkin (Physical Chemistry)*
*Birthday: 1910, Cairo, Egypt (British)*

Dorothy Hodgkin used x-ray crystallography techniques to determined the structure of penicillin, vitamin B12, and insulin. She was awarded the 1964 Nobel Prize in Chemistry for her determinations by X-ray techniques of the structures of important biochemical substances. A friend and advisor of Prime Minister Margaret Thatcher, Hodgkin was described as the cleverest woman in England, a gentle genius.

*Florence Nightingale*

# May 13

*Ronald Ross (Medicine)*
*Birthday: 1857, Almora, India*

Ronald Ross trained at Bartholomew's Medical School in London and joined the Indian Medical Service in 1881. He made a study of mosquitoes after learning they were believed to be carriers of malaria. His complex question was; which of several types of mosquito transmitted malaria, and of the many parasites in a mosquito which was the malarial parasite. He won a 1902 Nobel Prize in Physiology or Medicine for work on malaria, by which he has shown how it enters the organism and thereby has laid the foundation for successful research on this disease and methods of combating it.

# May 14

### Gabriel Fahrenheit (Physics)
### Birthday: 1686, Danzig

Gabriel Fahrenheit worked as a glass blower and physical instrument maker and by 1714 had made alcohol thermometers and mercury thermometers using various scales. He eventually settled on a scale of 0 to 212 where 0 degrees was the temperature of ice, water and salt, 32 of water and ice, and 212 of boiling water. The Fahrenheit scale is still a standard today.

### Charlotte Auerbach (Chemistry)
### Birthday: 1899, Crefeld, Germany

Charlotte Auerbach did research at Edinburgh using the fruit fly Drosophila and showed that mustard gas was a highly effective chemical mutagen. Although the results were a major contribution to science she was only acting as a technician in the discovery. She went on, however, to achieve a considerable body of work on the relation between chromosome breakage and gene mutation.

*Drosophila sp.*

# May 15

### Pierre Curie (Physics)
### Birthday: 1859, Paris, France

Pierre Curie, with his brother Jacques, discovered piezoelectricity in crystals of quartz under the influence of pressure. After his marriage to Marya Sklodovska in 1894, the two worked together in the study of radioactivity. Pierre concentrated on the physics and Marya (Marie) on the chemistry of the phenomenon.

# May 15

*Dorothy Andersen (Medicine, Pathology)*
*Birthday: 1901, Asherville, North Carolina, USA*

Dorothy Andersen, orphaned at 19, put herself through Mount Holyoke and Johns Hopkins and did medical research in two areas at Columbia. The first area was an extensive study of congenital heart problems. The second, for which she is better known, came from her 1935 discovery of cystic fibrosis. She developed diagnostic tests but no definitive treatment or cure.

# May 16

*Maria Gaetana Agnesi (Mathematics)*
*Birthday: 1718, Milan, Italy*

Maria Gaetana Agnesi was a child prodigy speaking six languages at age 11. Daughter of a mathematics professor at the University of Bologna, her chief love was mathematics and between the ages of 20 and 30 she wrote a two-volume textbook summarizing the algebra, geometry and calculus of her day. The book, *Le Instituzioni Analitiche* was published to great acclaim in 1748 and was still a useful text 50 years later.

*Ilya (Elie) Me(t)chnikov (Zoology)*
*Birthday: 1845, Kharkov, Eastern Ukraine*

Ilya Mechnikov studied the embryology of marine invertebrates at Kharkov University. Later he studied at Odessa (briefly) and in France with Pasteur, the involvement of phagocytes in digestion of food and bacteria, and senile atrophy. His theories are the earliest concept of auto-immune disease. He shared the 1908 Nobel Prize in Medicine or Physiology with Paul Ehrlich (May 29) in recognition of their work on immunity.

# May 17

*Edward Jenner (Medicine)*
*Birthday: 1749, Berkeley, Gloucestershire, England*

Edward Jenner was a pioneer of the science of immunology. He noticed that people who had had cowpox did not react to vaccinations with smallpox (practiced during outbreaks of smallpox.) By infecting a boy with cowpox and then smallpox, and then infecting with smallpox ten other people who had had cowpox, he showed that exposure to one provided immunity to the other.

# May 18

*Thomas Midgley (Chemistry)*
*Birthday: 1889, Beaver Falls, Pennsylvania, USA*

Thomas Midgley graduated from Cornell in mechanical engineering. He went to work for Delco where, in 1921, he discovered tetraethyl lead as an anti-knock additive for petrol engines. He later discovered that chlorofluorcarbons (freons) were good refrigerants.

# May 19

*Max Ferdinand Perutz (Molecular Biology)*
*Birthday: 1914, Vienna, Austria*

Max Perutz studied chemistry at the University of Vienna. He became interested in the work being done by biochemists at Cambridge and he went there in 1936 to determine the structure of the hemoglobin molecule using X-ray analysis. Perutz, with Kendrew (March 24), shared the 1962 Nobel Prize in Chemistry for his solution of the structure of hemoglobin.

*Hemoglobin Molecule*

## May 20

*Eduard Buchner (Chemistry)*
*Birthday: 1860, Munich, Bavaria, Germany*

Eduard Buchner's work on fermentation marked the beginning of modern enzyme chemistry. He ground yeast cells with sand and prepared a cell-free extract that he used to ferment the sugar sucrose to produce the alcohol ethanol. This showed that living yeast cells were not necessary (as had been believed) for fermentation. He was awarded the 1907 Nobel Prize in Chemistry for his biochemical researches and his discovery of cell free fermentation.

## May 21

*Lyon Playfair (Chemistry)*
*Birthday: 1818, Meerut, India*

Lyon Playfair was a British chemist who, like many other chemists of the time, qualified in medicine. He made some significant contributions in chemistry research, but was perhaps most important as an influential spokesman for advancing the cause of science and technology. He also worked on much needed educational reform in Britain.

## May 22

*William Sturgeon (Invention)*
*Birthday: 1783, Whittington, Lancashire, England*

William Sturgeon, an electrician and inventor, was self educated in languages and science. After leaving the army he set himself up as a bootmaker but also lectured in science at the Royal Military College. A practical man, he made the first workable electric motor, introduced the soft-iron electromagnet, invented the commutator, and worked on thermo electricity and on improving batteries.

# May 23

*Carl Linnaeus (Taxonomy)*
*Birthday: 1707, Rahult, Smaland, Sweden*

Carl Linnaeus was the son of Nils Ingemarsson. He adopted a new surname based on the name of the linden tree near his home. Linnaeus invented the binomial system of naming animals and plants with Latin names that still we use today. Every known animal or plant is classified with a generic name and a species name. He saw it as his mission to classify, record and name all living things.

*John Bardeen (Physics)*
*Birthday: 1908, Madison, Wisconsin, USA*

John Bardeen earned a BS in Electrical Engineering from Wisconsin and his PhD in mathematical physics at Harvard in 1935. In 1945 he went to work for the Bell Telephone Laboratory where he did the work for which he received the first of two Nobel Prizes. In 1951 he moved to the University of Illinois where he remained until his death in 1991. Bardeen is the only person so far to have been awarded two Nobel Prizes in Physics. The first, awarded in 1956 was shared with William Shockley (February 13) and Walter Brattain and was for the discovery of the transistor effect. The second, in 1972, was shared with Leon Cooper and was for their work on a theoretical explanation of superconductivity.

*Carl Linnaeus*

# May 24

*William Gilbert (Physics)*
*Birthday: 1544, Colchester, Essex, England*

William Gilbert trained in medicine and was physician to Queen Elizabeth I and James I of England. However, in science he is most remembered as an early investigator of magnetism. In his time the most powerful magnet known was the natural lodestone. Using this resource he discovered most of the properties of magnetism that we know today.

# May 25

*Pieter Zeeman (Physics)*
*Birthday: 1865, Zonnemaire, Netherlands*

Pieter Zeeman obtained his doctorate at Leiden, Investigating the effect of a magnetic field on light. Following up on some of Michael Faraday's (September 22) unsuccessful experiments, he found that a magnetic field widened the lines of a spectrum and named this the Zeeman Effect. The discovery led others to advances in spectroscopy. Zeeman shared the 1902 Nobel Prize in Physics for researches into the influences of magnetism upon radiation phenomena.

# May 26

*Sally Kristin Ride (Physics)*
*Birthday: 1951, Encino, California, USA*

Sally Ride received her Ph.D. from Stanford in physics, astronomy and astrophysics. She applied on a whim to NASA's astronaut training program and was accepted. She became the first American female astronaut to fly a space mission. She actually flew two missions – The Challenger STS-7 June 18-24, 1983 flight, and the STS-41G Challenger mission in 1984. She returned to academics as a professor of physics at the University of California, San Diego.

# May 27

*Rachel Louise Carson (Zoology)*
*Birthday: 1907, Springdale, Pennsylvania, USA*

Rachel Carson planned to be a writer but a first year class in biology at Pittsburgh changed her plans and she switched from English to zoology. She told a friend biology gave her something to write about. She taught zoology at Johns Hopkins and Maryland and then joined the US Bureau of Fisheries. Concerned about the world's global environment, she wrote radio scripts, articles and books and is best known for *Silent Spring* (1962).

## May 28

*Jean Louis Rodolphe Agassiz (Geology, Zoology, Natural History)*
*Birthday: 1807, Motier, Switzerland*

Jean Louis Agassiz had a lifelong interest in both fresh water and fossil fish. In 1846 he emigrated to America to study geology and natural history, and became professor of zoology and geology at Harvard. He published *The Natural History of the United States* in four illustrated volumes (1857-62). He believed not in Darwinian evolution but in successive independent creations.

*Wood Frog*

## May 29

*Chien-Shiun Wu (Physics)*
*Birthday: 1912, Liuhe, China*

Wu, a Chinese-born American, earned her doctorate in physics from the University of California, Berkeley. She earned an international reputation for her research in nuclear and particle physics by revising the existing understandings about the process of radioactive beta decay. She discovered that the particles in an atomic nucleus do not always behave symmetrically. Her discovery overturned the law of parity theory and stunned the physics community. The Nobel Prize however was awarded to the senior physicists Lee and Yang who had questioned the law of parity rather than to Wu and her colleague scientists who had experimentally disproved it. Her book *Beta Decay* published in 1966 is a standard reference book.

*Paul Ralph Ehrlich (Population Biology)*
*Birthday: 1932, Philadelphia, Pennsylvania, USA*

Paul Ralph Ehrlich took his doctorate at the University of Kansas. He is Bing Professor of Population Studies and Professor of Biology at Stanford. A well known and respected author he has written – often with his wife, Anne – a number of books concerning science and politics, and about his views on Earth's global environmental problems and on the need for social change in order to correct these problems.

# May 30

*Julius Axelrod (Pharmacology)*
*Birthday; 1912, New York, USA*

Julius Axelrod joined the staff of the National Institute of Mental Health in 1955. He shared the 1970 Nobel Prize in Medicine or Physiology with Bernard Katz (March 26) and Ulf von Euler (February 7) for work on the nervous system and the brain in which he had discovered the metabolic pathways for noradrenaline and adrenaline, isolated the enzymes involved, and explained the process.

# May 31

*Cagniard de Latour (Physics)*
*Birthday: 1777, Paris, France*

Latour trained at the Ecole Polytechnique in Paris. He contributed to science a succession of ingenious experiments. He invented, amongst other things, a perforated disc siren, a rotary machine for cleaning copper ore, and a dynamometer. His most significant discovery was of 'critical state' which is the point (characterized by temperature and pressure) at which the distinction between the gaseous and fluid phase of a liquid disappears.

*"Scientists study the world as it is,*
*engineers create the world that never has been."*

*Theodore von Karman (Hungarian/American Aeronautical Engineer)*

# June

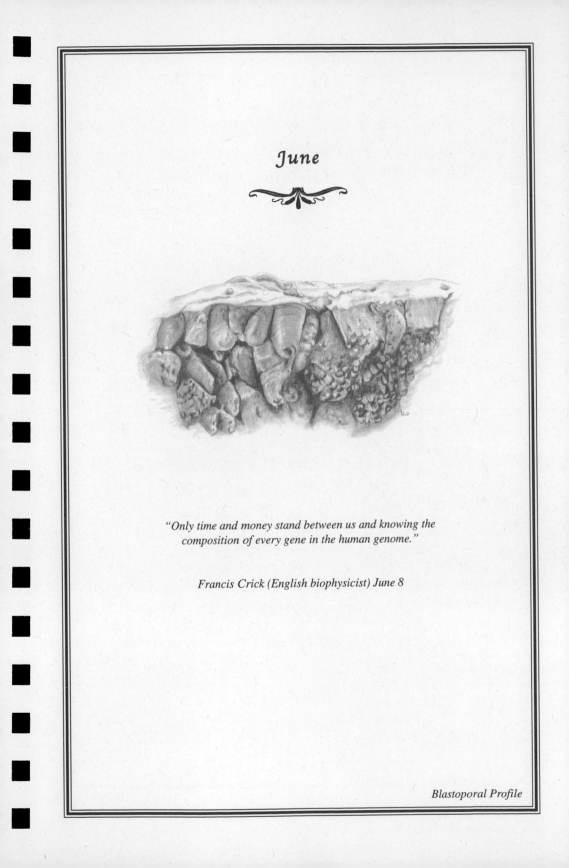

*"Only time and money stand between us and knowing the composition of every gene in the human genome."*

*Francis Crick (English biophysicist) June 8*

# Introduction to June

## *Learn Exciting Things and Go on to Accomplish Your Dreams*

" .... let me share a dream of mine with you.

Many years ago, before the Aqua-Lung, I used to dream a beautiful dream. I was flying in my dream, weightless and free, dipping and soaring and floating. It was a very happy feeling. When I started to dive on scuba under the surface of the sea, my dream stopped. Why? Because I was living my dream, wide-awake.

I have many waking dreams, though, things I try hard to accomplish. One of these dreams involves dolphins, and I would like to tell you why.

When dolphins are threatened by an animal of greater strength and size (a large shark, for instance), they come together. A pack of dolphins will suddenly form a tight group, dive below the shark, and drive their blunt noses into its belly, one after another. The shark is defeated by intelligence. The dolphins save themselves by joining together to do what they must.

You and I know we must save the health of our water planet and all creatures on earth. My dream is that, like dolphins, we can intelligently join together and work together to save our planet and ourselves. ...."

*Jacques-Yves Cousteau*
*December 6, 1984*

From a letter written by Jacques-Yves Cousteau to Explorit Science Center

# June

| | | | | |
|---|---|---|---|---|
| 1 | Frank Whittle, 1907 - 1996 | | 15 | Antoine Francoise de Fourcroy, 1755 - 1809 |
| 2 | George Henry Corliss, 1817 - 1888 | | 16 | Barbara McClintock, 1902 - 1992 |
| 3 | James Hutton, 1726 - 1797 | | 17 | Frederick John Vine, 1939 - |
| 4 | Jean Antoine C. Chaptal, 1756 - 1832 | | 18 | Charles Louis A. Laveran, 1845 - 1922 |
| 5 | Dennis Gabor, 1900 - 1979 | | | Jerome Karl, 1918 - |
| 6 | Edwin G. Krebs, 1918 - | | 19 | Blaise Pascal, 1623 - 1662 |
| 7 | James Young Simpson, 1811 - 1870 | | 20 | James Riddick Partington, 1886 - 1965 |
| 8 | Francis Harry Compton Crick, 1916 - | | 21 | Maximilian Wolf, 1863 - 1932 |
| 9 | George Stephenson, 1781 - 1848 | | 22 | William Macewen, 1848 - 1924 |
| 10 | Edward Penley Abraham, 1913 - | | 23 | Alan Mathison Turing, 1912 - 1954 |
| 11 | Jacques-Yves Cousteau, 1910 - 1997 | | 24 | Victor Franz Hess, 1883 - 1964 |
| | Keith Roberts Porter, 1912 - 1997 | | 25 | Herman Walther Nernst, 1864 - 1941 |
| 12 | Oliver Joseph Lodge, 1851 - 1940 | | 26 | W. (Lord Kelvin) Thomson, 1824 - 1907 |
| 13 | James Clerk Maxwell, 1831 - 1879 | | 27 | Hans Spemann, 1869 - 1941 |
| | Walter Luis Alvarez, 1911 - 1988 | | 28 | Maria Goeppert-Mayer, 1906 - 1972 |
| 14 | James Whyte Black, 1924 - | | 29 | George Ellery Hale, 1868 - 1938 |
| | | | 30 | Joseph Dalton Hooker, 1817 - 1911 |

# June 1

*Frank Whittle (Engineering)*
*Birthday: 1907, Coventry, Warwickshire, England*

Frank Whittle was in the airforce at the RAF College at Cranwell when he realized that the gas turbine being developed for industry, could be used to provide jet propulsion for aircraft. He took out his first patent for such an engine in 1930. A German Heinkel jet propelled plane flew in August 1939. The British Gloster with a Whittle jet engine flew in 1941 at 460 miles per hour.

# June 2

*George Henry Corliss (Engineering)*
*Birthday: 1817, Easton, New York, USA*

George Henry Corliss took out his first patent - for a mechanical shoe stitcher in 1842. He then became interested in steam engines and in 1856 opened the Corliss Engine Company. Corliss engines became very popular in cotton mills in Britain because they were smooth running, had sensitive controls and were economical.

# June 3

*James Hutton (Geology)*
*Birthday: 1726, Edinburgh, Scotland*

James Hutton trained in medicine but soon turned to agriculture. As a farmer in Berwickshire he became interested in the rocks below the soil he tilled and in 1795 published a two volume *Theory of the Earth* in which he laid the foundation of geology for the first time as an organized science.

# June 4

*Jean Antoine Claude Chaptal (Chemistry)*
*Birthday: 1756, Nogaret, Lozere, France*

Jean Antoine Claude Chaptal studied chemistry at Montpellier. His name is recognized in chaptalization, a process in which wine is enriched with sucrose (cane sugar). Chaptal was most interested in industrial aspects of chemistry and set up the first French factory to produce sulfuric acid. He also manufactured soda and white lead.

---

# June 5

*Dennis Gabor (Invention)*
*Birthday: 1900, Budapest, Hungary*

Dennis Gabor studied electrical engineering in Budapest but fled to Britain in the 1930s and worked for the Thomson-Houston Company. Gabor is remembered for the invention of holography which apparently occurred to him as he was waiting to play a game of tennis in 1947. He was working for the Imperial College of Science and Technology when he received the 1971 Nobel Prize in Physics for his invention and development of the holographic method.

*Skeletal Muscles*

# June 6

*Edwin G. Krebs (Biochemistry)*
*Birthday: 1918, Lansing, Iowa, USA*

Edwin G. Krebs studied the complex way in which proteins interact with cells specifically in connection with the mechanism of muscle contraction. The process that he revealed, in partnership with E.G. Fischer, has implications in relation to the immune response, the onset of cancer, and the release of hormones. Krebs and Fischer received a 1992 Nobel Prize for their discoveries concerning reversible protein phosphorylation as a biological regulatory mechanism.

## June 7

*James Young Simpson (Medicine)*
*Birthday: 1811, Bathgate, West Lothian, Scotland*

James Young Simpson studied medicine at Edinburgh and in 1847 was appointed Physician to the Queen in Scotland. Learning of experiments in France by Flourens using chloroform, Simpson started using this form of anesthesia in his obstetric practice. This was highly controversial until Queen Victoria chose to use chloroform in the birth in 1853 of her son Prince Leopold.

## June 8

*Francis Harry Compton Crick (Biophysics, Molecular Biology)*
*Birthday: 1916, Northampton, England*

Francis Harry Compton Crick was educated at University College London and Caius College Cambridge. He did research on three-dimensional structures of large biological molecules. With James Watson (April 6) and Maurice Wilkins he received the 1962 Nobel Prize in Physiology or Medicine for the discovery of the structure of DNA. He now works on the visual system of primates.

## June 9

*George Stephenson (Engineering)*
*Birthday: 1781, Wylam, Newcastle upon Tyne, England*

George Stephenson was a railway engineer, a pioneer of steam traction to replace horse drawn trains. In 1814 his steam engine Blucher ran on steel rails instead of wood. His engine Locomotion was the first public passenger train in the world and ran on the Stockton and Darlington Railway in 1825. In 1830 his famous Rocket engine ran at 30 mph on the new Liverpool Manchester Line.

# June 10

*Edward Penley Abraham (Biochemistry)*
*Birthday: 1913, Southampton, England*

Edward Penley Abraham worked on the chemistry of lysozyme, an antibiotic discovered in tears by Ambrose Fleming in 1921. He later worked on penicillin and his hypothesis about its structure was confirmed by Dorothy Hodgkins X-ray studies in 1945. In 1953, Abraham and Guy Newton isolated cephalosporin C, a new antibiotic.

*Banded butterflyfish*

# June 11

*Jacques-Yves Cousteau (Oceanography)*
*Birthday: 1910, St Andre-de-Cubzac, France*

Jacques-Yves Cousteau entered the French Naval Academy as a young man. During his rehabilitation after an automobile accident, he swam every day, learned to goggle-dive and fell in love with the underwater world. He developed a breathing regulator and other diving gear, and became world famous for his fascination with, and research into, the variety, interdependence and fragility of ocean life. His many books and documentary films helped to popularize the marine environment. Over years of diving and research, he saw the seas and the creatures in them changing for the worse, and founded The Cousteau Society (www.cousteau.org) with a mission to help protect the planet and its oceans.

*Keith Roberts Porter (Cell Biology)*
*Birthday: 1912, Yarmouth, NS Canada*

Keith Roberts Porter was a founding father of cell biology who pioneered the use of the electron microscope to examine biological cells. He received his doctorate in biology at Harvard and worked for many years at the Rockefeller Institute.

# June 12

*Oliver Joseph Lodge (Physics)*
*Birthday: 1851, Penkhull, Staffordshire, England*

Oliver Joseph Lodge became professor of physics and mathematics at Liverpool in 1881. He investigated the propagation of electromagnetic waves in wire and showed that hertzian waves could be used for telegraphic signals using Morse code. Lodge also investigated psychical phenomena.

---

# June 13

*James Clerk Maxwell (Physics)*
*Birthday: 1831, Edinburgh, Scotland*

James Clerk Maxwell studied at the Universities of Edinburgh, Cambridge, and London. In 1871 he became the first Cavendish Professor of Experimental Physics at Cambridge and was responsible for organizing the renowned Cavendish Labs. Maxwell is recognized as the creator of the electromagnetic theory of light but he also made contributions to the kinetic theory of gases.

*Walter Luis Alvarez (Physics)*
*Birthday: 1911, San Francisco, California, USA*

Walter Luis Alvarez worked with Ernest Lawrence at the University of California Radiation Laboratory at Berkeley. There he developed the hydrogen bubble chamber and pioneered the use of liquid hydrogen to make tracks of subatomic particles visible as trails of bubbles. He received the 1968 Nobel Prize in Physics for the discovery of a large number of resonance states.

# June 14

*James Whyte Black (Pharmacology)*
*Birthday: 1924, Cowdenbeath, Scotland*

James Whyte Black directed research at several pharmaceutical firms before becoming professor of analytical pharmacology at Kings College Medical School, London. With W. Duncan, Black pioneered the use of propanolol and other beta blockers for treatment of hypertension. He discovered the use of Tagamet in the treatment of ulcers and received a 1988 Nobel Prize in Physiology or Medicine for discoveries of important principles for drug treatment.

# June 15

*Antoine Francoise de Fourcroy (Chemistry)*
*Birthday: 1755, Paris, France*

Antoine Francoise de Fourcroy, with L.N. Vauquelin, showed that there were two series of mercury compounds. He developed the chemistry of urea and made many minor contributions to inorganic chemistry. But his most important contribution to chemistry was perhaps his advocacy of the views of Lavoisier (August 26).

# June 16

*Barbara McClintock (Genetics)*
*Birthday: 1902, Hartford, Connecticut, USA*

Barbara Mclintock was awarded the 1983 Nobel prize in Physiology or Medicine for her discovery of transposable genetic elements (jumping genes.) - a situation in which fragments of DNA can relocate themselves on chromosomes. She revolutionized maize (corn) genetics by mapping the position of the genes on its chromosomes.

*Barbara McClintock*

# June 17

### Frederick John Vine (Geology)
### Birthday: 1939, Brentford, Essex, England

John Frederick Vine was educated at London and at St Johns College, Cambridge. He became professor of environmental sciences at the University of East Anglia. With D.H. Matthews he developed the Vine-Matthews hypothesis confirming an earlier hypothesis concerning continental drift and seafloor spreading. This led to the theory of plate tectonics which revolutionized geology.

---

# June 18

### Charles Louis Alphonse Laveran (Medicine)
### Birthday: 1845, Paris, France

Charles Louis Alphonse Laveran was awarded the 1907 Nobel Prize in Physiology or Medicine for his work at the Pasteur Institute in Paris on the role played by protozoa in causing diseases. In November 1880 while studying black granules found in the blood of a malaria patient he had observed (for the first time) that flagella extruded from the granules and they moved. This confirmed his belief that malaria was caused by a protozoan parasite.

### Jerome Karl(Physical Chemistry)
### Birthday: 1918, Brooklyn, New York, USA

Jerome Karl researched the structure of atoms, molecules, glasses, crystals and solid surfaces. He is generally recognized for his X-ray crystallography studies of crystal molecules. For this work he shared the 1985 Nobel Prize in Chemistry for the development of direct methods for the determination of crystal structures.

*"When everything moves at the same time, nothing moves in appearance."*

*Blaise Pascal  (French philospher and mathematician) June 19*

## June 19

*Blaise Pascal (Mathematics)*
*Birthday: 1623, Clermont Ferrand, France*

Blaise Pascal did not have a formal education but by the time he was 18 he had done some of his best mathematical work. Before the age 16 he had proved an important conic theorem with his Pascal's Mystic Hexagram. He developed Pascal's Principle, and in 1665 Pascal's Triangle for calculating probabilities.

## June 20

*James Riddick Partington (Chemistry, History of Science)*
*Birthday: 1886, Bolton, Lancashire, England*

James Riddick Partington taught chemistry at London University. He was renowned for his encyclopedic knowledge. He is known now as a science historian. His works include: *Origins and Development of Applied Chemistry, A History of Chemistry*, his 4 volume *Advanced Treatise on Physical Chemistry* (1949-54) and finally, a *History of Greek Fire and Gunpowder* (1960).

## June 21

*Wolf Comet*

*Maximilian Wolf (Astronomy)*
*Birthday: 1863, Heidelberg, Germany*

Maximillian Wolf discovered in 1884 the comet that bears his name. He specialized in photographic astronomy and discovered several minor planets and nebulae. His work on nebulae was his most important. It was Wolf who established the presence of dark clouds of interstellar matter in our Milky Way Galaxy.

# June 22

*William Macewen (Medicine, Surgery)*
*Birthday: 1848, Rothesay, Scotland*

William Macewen was a student of Joseph Lister (April 5) and advocated and practiced aseptic surgery. In 1876 he designed an all steel osteome (no wood handle) so that it could be boiled. He boiled ligatures and needles, wore a sterilized gown, sterilized his patients skin and his own hands and the hands of his assistants and nurses. He pioneered brain, lung and orthopedic surgery.

# June 23

*Alan Mathison Turing (Mathematics)*
*Birthday: 1912, London, England*

Alan Mathison Turing was educated at Kings College, Cambridge. He worked on mathematical logic and developed a theoretical description of a universal computing machine (a Turing machine) to do any computation a human can do. He was a WW2 cryptologist and after the war worked at the National Physical Lab on the design of an automatic computing machine known as ACE. Turing's work is recognized as being significant in the developing area of computing machines.

# June 24

*Victor Franz Hess (Physics)*
*Birthday: 1883, Waldstein, Austria*

Victor Franz Hess was awarded the 1936 Nobel Prize in Physics for his discovery of cosmic radiation. He studied radioactivity and the source of background radiation. In a series of balloon ascents he measured ionization in sealed vessels at different altitudes and during a near total eclipse. He showed that the radiation came not from the sun but was cosmic radiation from outer space.

# June 25

*Herman Walther Nernst (Chemistry)*
*Birthday: 1864, Briessen, Germany*

Herman Walther Nernst's theories of galvanism are still basic to electrochemistry today. At the University of Berlin in 1905-6 he developed his Heat Theorem or Third Law of Thermodynamics - derived from earlier work by Helmholz (August 31). This theorem became connected with quantum theory and Nernst was awarded the 1920 Nobel Prize in Chemistry in recognition of his work in thermochemistry.

# June 26

*William (Lord Kelvin) Thomson (Physics)*
*Birthday: 1824, Belfast, Ireland*

William Thomson (Lord Kelvin) was professor of natural history in Glasgow when, with Michael Faraday, he developed the theory of electromagnetic fields. He played an important part in the laying of the first Atlantic cable. In 1851 he published *On the Dynamical Theory of Heat* in which he explained his absolute (Kelvin) scale of temperature.

*Blastoporal Profile*

# June 27

*Hans Spemann (Zoology, Embryology)*
*Birthday: 1869, Stuttgart, Germany*

Hans Spemann was a zoologist and skilled in the technique of microdissection. He showed that cells from one area of an embryo transplanted to another area would develop according to the transplant site unless the cells came from the blastopore when they would develop according to their origin. He won the 1935 Nobel Prize in Physiology or Medicine for discovering the blastopore organizer effect.

# June 28

*Maria Goeppert-Mayer (Mathematical Physics)*
*Birthday: 1906, Upper Silesia, Germany*

Maria Goeppert-Mayer entered Gottingen University to study mathematics but became fascinated by quantum mechanics. She later worked in the University of California, USA. She shared the 1963 Nobel Prize in Physics for the development of the atomic nuclear shell model theory that revolutionized the study of nuclear behavior.

# June 29

*George Ellery Hale (Astronomy)*
*Birthday: 1868, Chicago, Illinois, USA*

George Ellery Hale was the inventor in 1890 of the spectroheliograph. He founded the Astrophysical Journal and established a solar observatory at Mount Wilson (Pasadena, California). In 1928 he obtained funds from the Rockefeller Foundation for the 200 inch reflector at Mount Palomar. He discovered the 23 year cycle of reversal of polarity of the magnetic fields of sunspots.

# June 30

*Joseph Dalton Hooker (Botany)*
*Birthday: 1817, Halesworth, Suffolk, England*

Joseph Dalton Hooker was assistant surgeon and naturalist on the H.M.S. Erebus and Terror expedition to Antarctica in 1839-43. He later collected plants in Bengal, Nepal, Assam, Syria, Palestine, Morocco, and the Western US. He developed London's Kew Gardens (established by his father) into an international center for botanic research.

*"It was a great step in science when men became convinced that,
in order to understand the nature of things, they must begin by asking,
not whether a thing is good or bad, noxious or beneficial,
but of what kind it is? And how much is there of it?"*

*James Clerk Maxwell  (English physicist) June 13*

# July

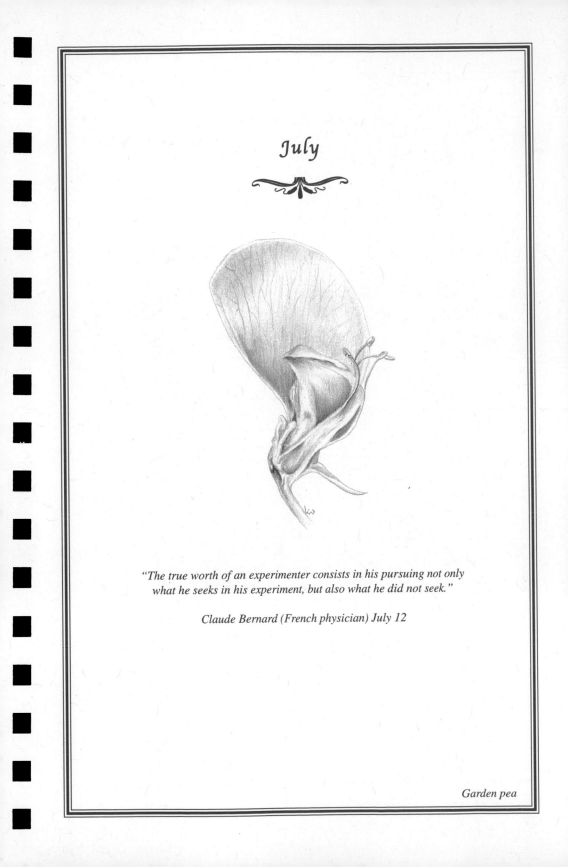

*"The true worth of an experimenter consists in his pursuing not only what he seeks in his experiment, but also what he did not seek."*

*Claude Bernard (French physician) July 12*

*Garden pea*

# Introduction to July

## The Delight in Creating Useful Mathematical Models

I became a scientist/engineer because of the delight I felt and still feel in creating mathematical models and understanding how they explain ordinary real world phenomena. One of the true joys of my life now is analyzing, often for the first time, how something works that no one else has understood before. My science may not be earthshaking in the sense that it will save millions of lives or significantly alleviate human suffering but it is joyous and useful. It is fulfilling to use one's mind to devise new mathematical models and uncover new understanding.

Sports have always fascinated me as well. All the spare pages in my boyhood schoolbooks were filled with drawings of football players, helmets, balls, and such. Because almost all sports involve motion of a participant, implements he or she uses, or vehicles of various sorts, they are therefore amenable to mathematical description using Newton's Laws of Motion. Sports also embody the idea of optimization, whether exceeding the performance of an opponent or just being the best that one can be. Mathematical models of sports can aid in our understanding of the events and lead to the development of training aids for competitors. The development and building of these machines and systems further appeals to the engineer in me.

Science is important to me in the larger view because it contains these two elements. Man masters his environment by understanding it. This understanding can often lead to new and useful machines or systems which improve our lives and make them more fun.

*Mont Hubbard*
*Sports Biomechanics Laboratory*
*Department of Mechanical Engineering, University of California, Davis*
*July 1997*

# July

| | | | |
|---|---|---|---|
| 1 | Ignaz Philipp Semmelweiss, 1818 - 1865 | 16 | Irmgard Flugge-Lotz 1903 1974 |
| 2 | Hans Albrecht Bethe, 1906 - | | Trinity Test 1945 |
| 3 | Soyuz 14, 1974 | 17 | Pierre L. M. de Maupertius 1698 - 1759 |
| 4 | Ivan Levinstein, 1845 - 1916 | 18 | Robert Hooke 1635 - 1703 |
| | Pathfinder 1997 | 19 | Rosalyn Sussman Yalow 1921 - |
| 5 | Andrew Ellicott Douglass, 1857 - 1962 | 20 | Edmund Percival Hillary 1919 - |
| 6 | Rudolph Albert Von Kolliker, 1817 - 1905 | | Apollo 11 1969 |
| 7 | Camillo Golgi, 1844 - 1926 | 21 | Jean Picard 1620 - 1682 |
| 8 | Alfred Binet, 1857 - 1911 | 22 | Johann Gregor Mendel 1822 - 1884 |
| 9 | Nikola Tesla, 1856 - 1943 | 23 | Vladimir Prelog 1906 - 1998 |
| 10 | Herbert Wayne Boyer, 1936 - | 24 | Charles Emil Picard 1856 - 1941 |
| 11 | Theodore Harold Maiman, 1927 - | | Ernst Heinkel 1888 - 1958 |
| | Sarah Blaffer Hrdy, 1946 | 25 | Rosalind Elsie Franklin 1920 - 1958 |
| 12 | Josiah Wedgewood, 1730 - 1795 | 26 | Carl Gustav Jung 1875 - 1961 |
| | Claude Bernard, 1813 - 1878 | 27 | Geoffrey de Havilland 1882 - 1965 |
| 13 | Marie Eugene Leon Freyssinet, 1879 - 1962 | 28 | Charles Hard Townes 1915 - |
| 14 | Jean Baptiste Andre Dumas, 1800 - 1884 | 29 | Charles William Beebe 1877 - 1962 |
| | Florence Bascom, 1862 1945 | 30 | Clive Marles Sinclair 1940 - |
| 15 | Jocelyn Bell Burnell, 1943 - | 31 | John Ericsson 1803 - 1899 |

# July 1

*Ignaz Philipp Semmelweiss (Medicine)*
*Birthday: 1818, Ofen, Hungary*

Ignaz Philipp Semmelweiss studied law, then medicine, and became Chair of Obstetrics at Budapest. When a friend died (1847) after pricking his finger during an autopsy, Semmelweiss was led to realize that infections could be carried from the sick to the healthy. He ordered all medical attendants to wash their hands in chloride of lime before surgery.

# July 2

*Hans Albrecht Bethe (Nuclear Physics)*
*Birthday: 1906, Strasburg, Alsace-Lorraine*

Hans Albrecht Bethe studied at Frankfurt and Munich. In 1933 he fled Nazi Germany and joined Cornel University in New York. He is probably best known for his work in 1938 on the energy produced by fusion of hydrogen into helium in the Sun and other stars. His book *Reviews of Modern Physics* has been called Bethe's Bible. He won the 1967 Nobel Prize in Physics for his contributions to the theory of nuclear reactions, especially his discoveries concerning the energy production in stars.

# July 3

*Soyuz 14*
*Event: 1974, Orbital flight*

The Soyuz 14 (Golden Eagle) Russian space craft was launched by the USSR on July 3, 1974. The craft carried two Russian cosmonauts, Pavel Popovich and Yuri Artyukhin, to rendevous with the space station Salyut 3. The mission was for Earth observation and military reconnaissance.

# July 4

*Ivan Levinstein (Chemistry)*
*Birthday: 1845, Charlottenberg, Germany*

Ivan Levinstein studied chemistry in Berlin. In 1864 he emigrated to England where, at age 19, he set up a factory to manufacture dyestuffs. The company later became the nucleus of the Dyestuffs division of ICI. Levinstein campaigned to reform British patent laws which he felt hampered the growth of the British Chemical Industry.

*Pathfinder*
*Event: 1997, Space Exploration*

The Pathfinder Mars Probe landed on Mars on July 4, 1997 after an eight month journey from Earth. This U.S. spacecraft landed surrounded by air bags on the dry, rocky surface of the planet Mars. A tiny rover called the Sojourner surveyed the landscape from within Pathfinder and transmitted pictures and other information back to Earth. On July 6th the rover was deployed and moved down a shallow ramp onto the actual surface. The radio transmissions took about 11 seconds to reach Earth.

# July 5

*Andrew Ellicott Douglass (Dendrochronology)*
*Birthday: 1857, Windsor, Vermont, USA*

Andrew Endicott Douglass held various university appointments in astronomy and made many astronomical observations. He invented a new science called dendrochronology - tree dating. Each year a tree increases its girth by adding a new cylinder of cells around its trunk. Each such annual ring is descriptive of the year's climate, and the number of rings indicates the age of the tree.

*Tree rings*

# July 6

### Rudolph Albert Von Kolliker (Biology)
### Birthday: 1817, Zurich, Switzerland

Rudolph Albert Von Kolliker studied medicine and science at Zurich. His biological work was wide ranging. He produced papers on histology, embryology, comparative anatomy and physiology, and Darwinism. He showed cleavage of cells in cuttlefish eggs; he showed a connection between nerves and nerve cells; and he studied the actions of such poisons as curare and strychnine.

*Golgi Apparatus*

# July 7

### Camillo Golgi (Histology)
### Birthday: 1844, Corteno, Italy

Camillo Golgi studied medicine at Pavia. In 1873 he devised a silver impregnation method of staining nervous tissue, and he classified nerve cells. In 1885 he differentiated between types of Malaria parasite. In 1898 he described the reticular Golgi apparatus in cell cytoplasm. He shared a 1906 Nobel Prize in Physiology or Medicine for work on the structure of the nervous system.

# July 8

### Alfred Binet (Psychology)
### Birthday: 1857, Nice, France

Alfred Binet went to Paris to study as a lawyer but, influenced by J.M. Charcot, instead studied neurology and psychology. His most notable contribution was his study of the development and measurement of intelligence. In 1905, with T. Simon, he published a series of mental tests to assess intellectual abilities and developmental norms in children. (The IQ test known as the Stanford-Binet was developed in the 1920's by Lewis Terman.)

# July 9

*Nikola Tesla (Electrical Engineering)*
*Birthday: 1856, Smitjan, Lika (former Jugoslavia)*

Nicola Tesla emigrated to the US in 1884. He worked for a few years with Thomas Edison (February 11) and then joined the Westinghouse Company. He developed the first alternating current induction motor and spent much time investigating the transmission of energy without wires. The Tesla Coil is a device that uses a transformer to produce a high-frequency high-voltage current. The device excites luminous discharges from certain gases.

---

# July 10

*Herbert Wayne Boyer (Genetics)*
*Birthday: 1936, Latrobe, Pennsylvania, USA*

Herbert Wayne Boyer studied bacteriology at Pittsburgh and then joined the Department of Genetics at UC San Francisco. In 1973 Boyer, in collaboration with Stanley Cohen of Stanford, showed that segments of two DNA molecules from different bacterial plasmid sources could be spliced and recombined. In 1976 he founded Genentech to exploit this technique of recombinant DNA.

---

# July 11

*Theodore Harold Maiman (Physics)*
*Birthday: 1927, Los Angeles, California, USA*

Theodore Harold Maiman studied at Columbia and Stanford before joining Hughes Research Laboratories to study masers (Microwave Amplification by Stimulated Emission of Radiation). In 1960 he succeeded in constructing the first optical maser or laser using a ruby cylinder with the ends ground flat and parallel.

# July 11

**Sarah Blaffer Hrdy (Anthropology)**
*Birthday: 1946, Dallas, Texas, USA*

Sarah Blaffer Hrdy studied at Wellesley and Radcliffe Colleges, and received her doctorate in anthropology from Harvard in 1975. Her field work in Rajasthan studying langurs (Indian monkeys) resulted in observations which challenged the prevailing beliefs regarding effects of overpopulation. It was generally accepted that infanticide in langur colonies was a response to the environmental stress of overpopulation. Hrdy discovered that infanticide by langur males happened even when the colonies were not overpopulated and she proposed an explanation based on evolutionary strategy benefiting individuals rather than a colony as a whole.

# July 12

*Wedgewood China*

**Josiah Wedgewood (Ceramics)**
*Birthday: 1730, Burslem, Staffordshire, England*

Josiah Wedgewood, born into a family of potters, sought always to improve the quality of his pottery. He was friendly with several noted scientists (Boulton, Keir, Priestly, Withering) and their advice about chemistry and mineralogy helped him devise new types of pottery such as green glaze, creamware, and jaspar. Wedgewood published papers on mineralogy and pyrometry.

**Claude Bernard (Medicine, Biology, Physiology)**
*Birthday: 1813, Saint-Juilien, France*

Claude Bernard qualified at Paris as a physician and then devoted himself to the experimental investigation of physiology. He showed that the stomach was not the entire seat of digestion, that certain nerves control the dilation of blood vessels, that red blood cells carry oxygen to body cells, that glycogen is stored in the liver, and showed how carbon monoxide acts as a poison.

*"If an idea presents itself to us, we must not reject it simply because it does not agree with the logical deductions of a reigning theory."*

Claude Bernard (French physician) July 12

# July 13

*Marie Eugene Leon Freyssinet (Civil Engineering)*
*Birthday: 1879, Objat, France*

Marie Eugene Leon Freyssinet had a talent for innovative design of reinforced concrete bridges and he invented the Freyssinet flat jack used in the midspan of many of his flat arched bridges. He designed some of the most elegant reinforced concrete bridges in the world and through his success, pre-stressed concrete became an important structural material.

# July 14

*Jean Baptiste Andre Dumas (Organic Chemistry)*
*Birthday: 1800, Alais, France*

Jean Baptiste Andre Dumas worked first as an apothecary apprentice and then went to Geneva where he concentrated on the investigation of iodine as a cure for goitre. Later, in Paris, he isolated anthracene from cola tar, and determined the constitution of chloroform. Then, in 1849, his determination of the atomic weight of carbon was a significant contribution to science.

*Florence Bascom (Geology)*
*Birthday: 1862, Williamstown, Massachusetts, USA*

Florence Bascom received her doctorate in 1893, the first woman to receive a Ph.D. from Johns Hopkins and the first woman to receive a Ph.D. in Geology in the U.S. She then spent most of her professional career at Bryn Mawr College where she founded and established the Geology Department. She also worked for the United States Geological Survey from 1896 mapping the crystalline rocks in southeastern Pennsylvania and adjoining states.

# July 15

*Jocelyn Bell Burnell (Astronomy and Physics)*
*Birthday: 1943, Belfast, Northern Ireland*

Jocelyn Bell Burnell discovered pulsars, a new class of dense, burned-out stars as a graduate student aged twenty four. The discovery opened up new areas of astronomy and provided clues to the evolution and death of stars. Her thesis advisor at Cambridge, Anthony Hewish, however, won the 1974 Nobel Prize in physics for his decisive role in the discovery of pulsars. Burnell herself defends this decision.

# July 16

*Irmgard Flugge-Lotz (Engineering)*
*Birthday: 1903, Hameln, Germany*

Irmgard Flugge-Lotz earned her doctorate in engineering in Germany in 1929. In 1931 she developed a mathematical formula for determining the distribution of lift over the span of an aircraft's wings. She immigrated to the United States in 1948 and worked as a lecturer and research supervisor at Stanford University where she perfected research on automatic aircraft controls that she had been working on in Germany. in 1960 Lotz became Stanford's first woman professor of engineering.

*Trinity Test*
*Event: 1945, Alamogordo, New Mexico, USA*

The Trinity Test happened in the predawn of Monday, July 16, 1945. A new world era began at 5:29 a.m. when a 10,000 pound atom bomb was detonated in the desert at the New Mexico Trinity Site. The resulting flash was seen 250 miles away and a mushroom cloud rose high in the sky over the desert. A duplicate bomb destroyed Nagasaki, Japan twenty-four days later.

# July 17

*Pierre Louis Moreau de Maupertius (Mathematics, Geodesy)*
*Birthday: 1698, St Malo, France*

Pierre Louis Moreau de Maupertius taught physics for several years at the Berlin Academy but spent later years in France. He led an expedition in 1731 to Lapland to measure the exact length of a meridian degree. The results, confirming a prediction by Newton (December 25), showed that the earth is not a perfect sphere. Voltaire called Maupertius The Great Flattener.

# July 18

*Robert Hooke (Physics)*
*Birthday: 1635, Isle of Wight, England*

Robert Hooke, a brilliant and versatile scientist, started his career as an assistant to Robert Boyle (January 25) in Oxford. Later, as professor of geology, he studied in some of the same areas as Newton. As an inventor and designer, however, he was unrivaled. He invented a spring control for a watch balance wheel, the compound microscope, a wheel barometer, and astronomical instruments.

*Hooke's Microscope*

# July 19

*Rosalyn Sussman Yalow (Medical Physics)*
*Birthday: 1921, New York City, USA*

Rosalyn Sussman Yalow was interested in endocrinology, especially in relation to diabetes. She was awarded the 1977 Nobel Prize in Physiology or Medicine for the development of the technique of radioimmunoassay (a spectacular combination of immunology, isotope research, mathematics and physics) which can be used in the diagnosis and treatment of various diseases. It can be used for example to assay peptide hormones such as insulin.

# July 20

*Edmund Percival Hillary (Mountaineering)*
*Birthday: 1919, Auckland, New Zealand*

Edmund Percival Hillary, a New Zealand explorer and mountaineer worked as an apiarist with his brother between expeditions. His first expedition, in 1951, was the Garwhal Expedition to the Himalayas. He then joined the Everest Expeditions and, with Norgay Tenzing, reached the Everest peak on May 29, 1953. Since then he has led expeditions to many other places and has written about his exploits.

*Apollo 11*
*Event: 1969, US Space Program*

Neil Armstrong (August 5), the first human being to step onto the Moon, took his lunar stroll on July 20, 1969. This achievement was part of the Apollo 11 mission of the US Space Program. A lunar module – the "Eagle" – landed both Armstrong and Buzz Aldrin near the Sea of Tranquility. As he stepped onto the Moon surface, Armstrong said, "One small step for a man, one giant leap for mankind."

# July 21

*Jean Picard (Astronomy)*
*Birthday: 1620, La Fleche, France*

Jean Picard made telescopic sights for astronomical instruments, and improved William Gascoigne's micrometer. Then, after ten years of observations, noted that the Pole Star varies its position by up to forty seconds of arc annually. An important contribution was his measurement of the length of a meridian degree which he found to be 69.104 miles, a fact later used by Newton (December 25).

# July 22

### Johann Gregor Mendel (Genetics)
### Birthday: 1822, Heinzendorf, Silesia

Johann Gregor Mendel entered the Augustinian Order of monks in 1843. With an interest in evolution and skill as a gardener, he spent many years growing thousands of pea plants that he meticulously cross-pollinated by hand in order to elucidate the role of hybridization in evolution. His resulting theories were sound and his work laid the foundation of the science of genetics. However, the work was not published until 1900 and Charles Darwin (July 22) developed his evolutionary theories without the benefit of knowing about Mendel's work.

*Garden pea*

# July 23

### Vladimir Prelog (Chemistry)
### Birthday: 1906, Sarajevo, Bosnia Herzegovina

Vladimir Prelog trained as a scientist in Prague and then worked as an industrial chemist. In 1941 he fled to Switzerland. Here he did structural and synthetic studies of terpenes, steroids, alkaloids, and antibiotics. In 1967 he synthesized the antibiotic boromycin. He shared the 1975 Nobel Prize in Chemistry for his research into the stereochemistry of organic molecules and reactions.

# July 24

### Charles Emil Picard (Mathematics)
### Birthday: 1856, Paris, France

Charles Emil Picard was professor of mathematics at Toulouse and then at the Sourbonne. His contributions to mathematics were in the field of analysis and analytic geometry. He used successive approximations to show solutions of ordinary differential equations. he contributed to the development of algebraic geometry. He applied analysis to the investigation of elasticity, heat and electricity.

# July 24

*Ernst Heinkel (Engineering)*
*Birthday: 1888, Grunbach, Swabia, Germany*

Ernst Heinkel was a gifted airplane designer and built his first plane in 1911. During WW1 he designed the W-12 sea plane and the W-20 monoplane. He continued designing planes between the two world wars and by 1939 had designed and built the Heinkel He-100 (the fastest plane in the world at the time) which he sold to the Russians. In collaboration with Werner von Braun he designed the He-176, the world's first rocket plane, and then the He-178, the first jet plane.

*Rosalind Franklin*

# July 25

*Rosalind Elsie Franklin (X-ray Crystallography)*
*Birthday: 1920, London, England*

Rosalind Elsie Franklin studied chemistry at Newnham College, Cambridge. She became expert at taking X-ray photographs and used X-ray crystallography to establish in 1951 that DNA exists in two forms (A and B). By early 1953 she realized that her photograph of the B-form showed a two-chained helix. Influenced by their knowledge of that photograph, James Watson (April 6) and Francis Crick (June 8) determined that the DNA molecule was a double helix with paired bases.

# July 26

*Carl Gustav Jung (Psychology)*
*Birthday: 1875, Kesswil, Switzerland*

Carl Gustav Jung studied medicine at Basle and then turned to psychiatry. He was the founder of analytical psychology. He met Sigmund Freud (May 6) in 1907 and they worked closely for several years. Jung turned to anthropological material to develop his ideas about universal symbolism and departed from Freud's ideas which he then found incomplete. He published *Psychological Types* in 1923.

# July 27

*Geoffrey de Havilland (Engineering, Aeronautics)*
*Birthday: 1882, High Wycombe, England*

Geoffrey de Havilland was educated as a mechanical engineer. He designed his first plane and engine in 1908-9 and taught himself to fly in 1910. Of the eight planes he designed, five went into war service. His Moth (1925) became the Tiger Moth, a standard WW2 trainer. He developed the Mosquito, and the Comet which was the first passenger jet airliner.

# July 28

*Charles Hard Townes (Physics)*
*Birthday: 1915, Greenville, South Carolina, USA*

Charles Hard Townes worked at the Bell Telephone Laboratories during WW2. Later he was one of three scientists who independently developed the technique of microwave spectroscopy of gasses. By 1953 he had constructed the first maser. He envisioned it first as a way to study atoms and molecules but it found applications too in astronomy. Townes shared the 1964 Nobel Prize in Physics for fundamental work in the field of quantum electronics which has led to the construction of oscillators and amplifiers based on the maser-laser principle.

# July 29

*Charles William Beebe (Natural History)*
*Birthday: 1877, Brooklyn, New York, USA*

Charles William Beebe graduated from Columbia and worked at the New York Zoological Gardens. He was very interested in Ornithology but is perhaps better remembered as a pioneer explorer of deep sea natural history. In his Bathysphere with Otis Barton, he reached a depth of 1000 meters near Bermuda in 1934 and later reached a depth of close to a mile.

# July 30

*Clive Marles Sinclair (Invention)*
*Birthday: 1940, Richmond, Surrey, England*

Clive Marles Sinclair has been dubbed the Heath Robinson of our time based on the multitudes of his electronic inventions. A pocket calculator in 1972 was followed by other calculators, a watch, a pocket TV, and then a series of personal computers culminating in the ZX Spectrum in 1982. Sinclair's computers played a large part in the microcomputing boom of the 1980s.

# July 31

*John Ericsson (Invention)*
*Birthday: 1803, Farnebo, Sweden*

John Ericsson produced a new invention from 1826 each year for twelve years. His inventions included a steam locomotive (in competition with Stephenson's Rocket), and a steam frigate with a screw propeller. In 1848 he became an American citizen and built the turreted, ironclad, Civil War vessel *Monitor* using plans he had earlier offered to Napoleon III.

*Sinclair calculator*

*"Among scientists are collectors, classifiers and compulsive tidiers-up;
many are detectives by temperament and many are explorers, some are
artists and others artisans. There are poet-scientists and philosopher-scientists
and even a few mystics...and most people who are in fact scientists
could easily have been something else instead."*

*Peter B. Medawar (English biologist) February 28*

# August

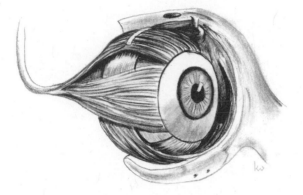

"Any pride I might have held in my conclusions was perceptibly
lessened by the fact that I knew that the solution of these problems
had almost always come to me as the gradual generalization of
favorable examples, by a series of fortunate conjectures, after many errors."

*Herman von Helmholtz (German physiologist) August 31*

*Human eye*

# Introduction to August

## Influences and the Drive to Scientific "Firsts"

I embarked on a career in science from an innate desire to understand how the human body works. I also found that science was able to satisfy my desire to tinker with things. Growing up in Brooklyn, N.Y., my early days in high school were far from academic. My scientific desires were first unearthed and initially crystallized by my 10th grade biology teacher. As in my case, a good teacher can "make or break" your interests in a particular subject. Another important influence in my scientific career was an uncle who was a physical chemist and entrepreneur.

After completing my Ph.D. in biochemistry, I had the good fortune of "cutting my scientific teeth" as a molecular biology Postdoctoral Fellow at Genentech, Inc. from 1982-1984. These were the fast-paced early days of genetic engineering and recombinant DNA technology. We were creating and using the technology as we went along. Working 80 hours per week in the lab was not uncommon. The sheer excitement of the potential to be the first person in the world to clone and express a human gene made the work intensity a pleasure. It was like "playing in the Superbowl."

Being a key member of a team of scientists that was the first to clone, express and characterize human tumor necrosis factors (TNF) alpha and beta, now two of the most widely studied human immune system regulatory proteins, was a thrill that has given me a secure sense of scientific achievement. It is very satisfying to have seen others follow up on TNF research, which has led to an inhibitor that is an approved therapeutic treatment for Rheumatoid Arthritis.

The excitement of being able to achieve scientific "firsts" is still a factor in what motivates my work in industrial enzymology today.

*Glenn E. Nedwin, Ph.D.*
*President*
*Novo Nordisk Biotech, Inc.*
*December 1999*

# August

| | | | |
|---|---|---|---|
| 1 | Jean Baptiste Lamarck, 1744 - 1829 | 19 | John Flamsteed, 1646 - 1719 |
| 2 | John Tyndall, 1820 - 1893 | | Orville Wright, 1871 - 1948 |
| 3 | Elisha Graves Otis, 1811 - 1861 | 20 | Roger Wolcott Sperry, 1913 - |
| 4 | William Rowan Hamilton, 1805 - 1865 | 21 | William Kelly, 1811 - 1888 |
| 5 | Neil Alden Armstrong, 1930 - | 22 | Denis Papin, 1647 - 1712 |
| | Shirley Ann Jackson, 1946 - | 23 | George Leopold Cuvier, 1769 - 1832 |
| 6 | Alexander Fleming, 1881 - 1955 | | Estelle Ramey, 1917 - |
| 7 | Louis Seymour Bazett Leakey, 1903 - 1972 | 24 | Mount Vesuvius, 79 |
| 8 | Florence Merriam Bailey, 1863 - 1948 | | Elizabeth Lee Hazen, 1885 - 1975 |
| 9 | Thomas Telford, 1757 - 1834 | 25 | Hans Adolph Krebs, 1900 - 1981 |
| 10 | Arne Wilhelm K. Tiselius, 1902 - 1971 | 26 | Antoine Laurent Lavoisier, 1743 - 1794 |
| 11 | Erwin Chargaff, 1905 - | 27 | Flossie Wong-Staal, 1946 - |
| 12 | Erwin Schrodinger, 1887 - 1961 | | Marina 2, 1962 |
| 13 | John Logie Baird, 1888 - 1946 | 28 | Godfrey Newbold Hounsfield, 1919 - |
| 14 | Charles Naudin, 1815 - 1899 | 29 | Oliver Wendell Holmes, 1809 - 1894 |
| 15 | Gerty Radnitz Cori, 1896 - 1957 | 30 | Ernest Rutherford, 1871 - 1937 |
| 16 | Arthur Cayley, 1821 - 1895 | | Edward Mills Purcell, 1912 - 1997 |
| 17 | Pierre de Fermat, 1601 - 1665 | 31 | Michel Eugene Chevreul, 1786 - 1889 |
| 18 | Brook Taylor, 1685 - 1731 | | Herman Helmholtz, 1821 - 1894 |

# August 1

*Jean Baptiste Lamarck (Natural History)*
*Birthday: 1744, Paris, France*

Jean Baptiste Lamarck studied medicine in Paris until his interest turned to botany. In 1778 he published *Flore Francaise,* a dichotomous botanical flora (key). In 1793 he turned his attention to the animal kingdom and invertebrates. He developed four laws to explain evolution. Three of these are compatible with Darwinian (February 12) theories, but the fourth concerns inheritance of acquired traits.

# August 2

*John Tyndall (Physics)*
*Birthday: 1820, Leighlin Bridge, Ireland*

John Tyndall studied chemistry under Robert Bunsen (March 31) and his most important work was on heat. He is also remembered by the Tyndall effect in which light is scattered by fine particles in the air. Tyndall was one of the first to climb the Matterhorn and made important contributions to glacier theories. He also confirmed the impossibility of spontaneous generation - as did Pasteur (December 27).

# August 3

*Elisha Graves Otis (Invention)*
*Birthday: 1811, Halifax, Vermont, USA*

Elisha Graves Otis was a master mechanic for a bed-stead maker. He was put in charge of the construction of the company's new factory at Yonkers, New York. He designed a safety elevator for the building and gave a sensational demonstration of its effectiveness at a New York fair in 1854. He invented a steam elevator which was commercialized successfully by his sons.

# August 4

*William Rowan Hamilton (Physics, Mathematics)*
*Birthday: 1805, Dublin, Ireland*

William Rowen Hamilton knew 13 languages by age 13 and had a great facility for complex mental computation. He studied at Trinity College Dublin and was appointed professor of astronomy before completing his degree. His contributions to science were in optics, and in mathematics on which he concentrated as a recluse in later years.

*Moonprint*

# August 5

*Neil Alden Armstrong (Aeronautics, Space Exploration)*
*Birthday: 1930, Wapakoneta, Ohio, USA*

Neil Alden Armstrong entered Purdue in 1947 to study aeronautical engineering. He later became a civilian test pilot and in 1962 began training as an astronaut. He was one of three members of the Apollo 11 Mission (July 20) and was the first person to set a foot on the Moon surface on July 20, 1969. He later became a professor of aerospace engineering at Cincinnati.

*Shirley Ann Jackson (Physics)*
*Birthday: 1946, Washington, D.C., USA*

Shirley Ann Jackson received her B.S. from Massachusetts Institute of Technology in 1968 and her Ph.D. (Physics) in 1973. Jackson became the first African American female to receive a doctorate in Theoretical Solid State physics from MIT. She later switched fields to theoretical condensed matter physics.

# August 6

*Alexander Fleming (Medicine)*
*Birthday: 1881, Lochfield, Scotland*

Alexander Fleming studied and then worked at St Mary's Hospital Medical School in London. He is remembered for two discoveries. In 1922 he discovered lysozyme in tears and nasal secretions and showed that it dissolved certain bacteria. Alerted by a 1928 chance observation, he isolated the mold Penicillium notatum and showed that its antibacterial property had no effect on white blood cells.

# August 7

*Louis Seymour Bazett Leakey (Anthropology)*
*Birthday: 1903, Kabele, Kenya*

Louis Seymour Bazett Leakey was the son of missionaries in Africa. He was educated at Cambridge and received a first class degree in anthropology and archeology. He is renowned for his discoveries of early hominid fossils in East Africa. His excavations in the Olduvai Gorge revealed hominid remains that both clarified and confused the evolutionary tree of *Homo sapiens sapiens*.

# August 8

*Florence Merriam Bailey (Ornithology)*
*Birthday: 1863, Locust Grove, New York, USA*

Florence Augusta Merriam Bailey planned to be a writer and attended Smith College in Massachusetts. She developed a serious interest in natural history and became an expert ornithologist. She wrote books for young people and standard reference texts. Her first reference text was *Merriam's Handbook of Birds of the Western United States* (1902).

# August 9

*Thomas Telford (Civil Engineering)*
*Birthday: 1757, Westerkirk, Eskdale, Scotland*

Thomas Telford designed and supervised the construction of roads, bridges, aqueducts and canals. He supervised the construction of the Gotha canal in Sweden, and 920 miles of road and 1200 bridges in the Highlands of Scotland. He used iron beds for some of his canals and build five cast-iron road bridges.

# August 10

*Arne Wilhelm Kaurin Tiselius (Biochemistry)*
*Birthday: 1902, Stockholm, Sweden*

Arne Wilhelm Kaurin Tiselius studied at Uppsala and became professor there of biochemistry. He soon became interested in electrophoresis of proteins and introduced several improvements in the techniques of the time. He also studied adsorption and developed some general techniques of chromatography. He was awarded the 1948 Nobel Prize in Chemistry especially for his discoveries concerning the complex nature of the serum proteins.

*Human Serum Albumin*

# August 11

*Erwin Chargaff (Biochemistry)*
*Birthday: 1905, Czernowitz, Bohemia (now Czech Republic)*

Erwin Chargaff's initial work was on the coagulation of blood. Turning to the study of nucleic acids he showed that DNA bases are in complementary pairs (Chargaff's Rules). This information was crucial to James Watson (April 6) and Francis Crick (June 8) in their determination of the structure of the DNA double helix. In later years Chargaff became critical of science and pessimistic about man's future.

# August 12

*Erwin Schrodinger (Physics)*
*Birthday: 1887, Vienna, Austria*

Erwin Schrodinger studied in Vienna. His most important work was done in Zurich where he published scientific papers that were the foundation of wave mechanics. He took the earlier theories of Broglie and Einstein and confirmed them mathematically. The result when combined with the work of Bohr, Dirac, and Pauli became Quantum Mechanics.

# August 13

*John Logie Baird (Electrical Engineering)*
*Birthday: 1888, Helensburg, Scotland*

John Logie Baird was an electrical engineer but, plagued by ill health, retired to live at Hastings in 1922. He turned then to amateur experimentation and in 1924 had contrived a primitive television apparatus capable of transmitting a picture over several feet. By 1928 he was able to transmit and receive between London and New York. He went on to pioneer color and stereoscopic TV.

# August 14

*Charles Naudin (Botany)*
*Birthday: 1815, Autun, France*

Charles Naudin taught botany in France in 1844. Two years later he became stone deaf and gave up teaching to do taxonomic studies. In 1852 he suggested that hybridization might play a part in evolution. Between 1854 and 1864 he grew several generations of hybrids to test his theories but his techniques were inferior to the still generally unknown work by Gregor Mendel (July 22) and his results were ultimately less useful.

## August 15

*Gerty Radnitz Cori (Biochemistry)*
*Birthday: 1896, Prague, Czechoslovakia*

Gerty Radnitz Cori was the first American woman to win a Nobel Prize in science. With her husband Carl she was a pioneer in the study of enzymes and hormones. The Cori team explained the cycle of carbohydrates in the body - known as the Cori cycle. They were awarded the 1947 Nobel prize in Physiology or Medicine for their discovery of the course of the catalytic conversion of glycogen.

*Gerty Radnitz Cori*

## August 16

*Arthur Cayley (Mathematics)*
*Birthday: 1821, Richmond, Surrey, England*

Arthur Cayley graduated from Trinity College, Cambridge, and worked in law for fourteen years before returning to Cambridge as professor of mathematics. Cayley can lay claim to being the inventor of the theory of matrices. He also broke ground in the area of abstract geometry and the analytical geometry of curves. His output of published work was considerable.

## August 17

*Pierre de Fermat (Mathematics)*
*Birthday: 1601, Beaumont-de-Lomagne, France*

Pierre de Fermat was an amateur mathematician. In his other life he was a civil servant (Counsellor in the Parlement de Toulouse), classical scholar and poet. Some of his theorems were presented to the world as challenges to other mathematicians. He was interested in the theory of numbers, and with Pascal (June 19), can lay claim to being a founder of probability theory.

# August 18

**Brook Taylor (Mathematics)**
**Birthday: 1685, Edmonton, London, England**

Brook Taylor graduated from Cambridge and his friends included Isaac Newton (December 25) and Edmund Halley (October 29). He wrote several papers on physical subjects. His mathematical writings included *Methodus Incrementorium Directa et Inverta* which explained his Taylor's Series. He was a gifted musician and artist and applied geometrical methods to perspective in his art.

---

# August 19

**John Flamsteed (Astronomy)**
**Birthday: 1646, Derby, England**

John Flamsteed entered Jesus College Cambridge in 1670 to study astronomy. In 1675 he became the first Astronomer Royal at the new Royal Observatory, Greenwich. The observatory was ill equipped at the time but Flamsteed persevered with what instruments he was able to obtain. He achieved about 20,000 observations and published a British catalogue of 2884 stars.

*Wright Brothers*

**Orville Wright (Invention)**
**Birthday: 1871, Dayton Ohio**

Orville Wright, with his brother Wilbur, designed and constructed the first successful airplane. In 1900 and 1901 the brothers tested their first gliders remaining airborne each time for less than 60 seconds.. By 1905 they had experimented with planes powered by a petrol engine and in October 1905 they flew a powered aircraft on a 24 mile circuit. They patented their invention in 1906.

*"The wildest stretch of imagination of that time would not have permitted us to believe that within a space of fifteen years actually thousands of these machines would be in the air engaged in deadly combat."*

*Orville Wright (American aviation inventor) August 19*

# August 20

*Roger Wolcott Sperry (Zoology)*
*Birthday: 1913, Hartford, Connecticut, USA*

Roger Wolcott Sperry studied zoology at Oberlin and Chicago. He was interested in nervou
regeneration in lower animals, and in the functions of the two halves (right and left) of th
brain in higher animals. In humans he found the left brain to control speech, and the righ
brain non-verbal processes. He shared a 1981 Nobel Prize in Physiology or Medicine fo
his discoveries concerning the functional specialization of the cerebral hemispheres

# August 21

*William Kelly (Invention)*
*Birthday: 1811, Pittsburgh, Pennsylvania, USA*

William Kelley owned land containing iron ore where he worked a Cobb furnace. He also
had factories for making sugar-boiling kettles where pig iron was converted to wrought
iron. Attempting to make the converter process more economical he developed a way of
removing excess carbon by burning it out of molten iron with an air blast. He obtained a
patent for this process in 1857.

# August 22

*Denis Papin (Engineering)*
*Birthday: 1647, Blois, Loire et Cher, France*

Denis Papin was an assistant to Christiaan Huygens (April 14), then to Robert Boyle (January
25), and eventually went to Marburg as professor of mathematics. He invented a digester
which boiled at 16 atmospheres per square inch (psi) at a temperature of 200 degrees
centigrade. He also used steam as a source of power and his method was later used by
Newcomen who made the first successful steam engine.

# August 23

*George Leopold Cuvier (Natural History)*
*Birthday: 1769, Montbeliard, France*

George Leopold Cuvier was educated at the Academy of Stuttgart. He became assistant to the professor in comparative anatomy at the Natural History Museum in Paris, and eventually was professor there. He was interested in the structure and classification of animals including fossil mammals and reptiles. He laid the foundation for modern paleontology.

*Estelle Ramey (Physiology)*
*Birthday: 1917, Detroit, Michigan, USA*

Estelle Ramey was an endocrinologist. Her research at the University of Chicago School of Medicine and the Georgetown University Medical School focused mainly on hormone systems including insulin and cortisone. She earned a BS in biology in 1926 at age 19, an M.A. in physical chemistry in 1941, and doctorate in physiology and endocrinology in 1950.

# August 24

*Elizabeth Lee Hazen (Medicine)*
*Birthday: 1885, Rich, Mississippi, USA*

Elizabeth Lee Hazen earned her doctorate at Columbia University at the age of 42 having already established herself as a bacteriologist at a West Virginia Hospital. She became expert at identifying disease-causing fungi and worked to find substances that could be used effectively in the treatment of such diseases. Working with chemist Rachel Brown she announced in 1950 their discovery of a microbe that produced fungus-killing substances. The drug produced as a result of this discovery was named Nystatin and was marketed in 1954 and has proved to have a surprising range of human and agricultural uses.

*"The observer listens to nature:*
*the experimenter questions and forces her to reveal herself."*

*Georges Cuvier (French naturalist) August 23*

# August 24

*Mount Vesuvius*
*Event: 79, Italy*

On August 24, 79 AD, earthquakes which had been growing more violent in the area about 7 miles east of Naples, Italy, culminated in a tremendous volcanic eruption from Mount Vesuvius. Three towns were destroyed and more than 2,000 people died, many of them suffocated by the poisonous fumes.

*Mt. Vesuvius*

# August 25

*Hans Adolph Krebs (Biochemistry)*
*Birthday: 1900, Hildesheim, Germany*

Hans Adolph Krebs studied medicine and did biochemical research under Otto Warburg in Berlin. With the help of the Rockefeller Foundation in the 1930s at Cambridge and then Oxford, he carried out research on metabolic pathways which resulted in the development of the cycle of urea synthesis in the liver, and the tricarboxylic acid Krebs Cycle. He shared a 1953 Nobel Prize in Physiology or Medicine for his discovery of the citric acid cycle.

# August 26

*Antoine Laurent Lavoisier (Chemistry)*
*Birthday: 1743, Paris, France*

Antoine Laurent Lavoisier studied law in Paris. He also studied astronomy, botany, chemistry, geology and mathematics. His wife, Marie Paulze, assisted him in his wide ranging researches. Of great import was his explanation of combustion, not as the liberation of phlogiston as had been believed for more than a century, but as the result of combination with oxygen.

# August 27

## Flossie Wong-Staal (Genetics)
### Birthday: 1946, Kuangchou, China

Flossie Wong-Staal attended school in Hong Kong and college at the University of California, Los Angeles where she earned her Ph.D. in 1972. Wong-Staal then joined the National Institutes of Health in Bethesda Maryland and worked in the laboratory of Robert Gallo (credited with discovering the AIDS virus). Wong-Staal was the first scientist to clone the AIDS virus genes. She established the sequencing of the genes and their functions. Moving in 1990 to San Diego, California, Wong-Staal continues her research focusing on AIDS treatment research.

## Mariner 2
### Event: 1962, Space Exploration

Mariner 2, the first successful interplanetary spacecraft, was launched by the US on August 27, 1962. It passed within 21,000 miles of Venus on December 14, 1962 and recorded the temperature of the Venusian atmosphere as being about 500 degrees Celsius (900 degrees Fahrenheit.)

*Mariner spacecraft*

# August 28

## Godfrey Newbold Hounsfield (Electrical Engineering)
### Birthday: 1919, Newark, England

Godfrey Newbold Hounsfield served in the RAF in WW2 and then joined Electrical and Musical Industries (EMI) where he led the design team that built the UKs first large, solid state computer, EMIDEC 1100 (1959). He then developed the first Computerized Axial Tomography (CAT) machine for medical X-ray diagnosis. He shared the 1979 Nobel Prize in Physiology or Medicine with Alan M. Cormack (February 23) for the development of computer assisted tomography.

# August 29

*Oliver Wendell Holmes (Medicine)*
*Birthday: 1809, Cambridge, Massachusetts, USA*

Oliver Wendell Holmes entered Harvard to study law but switched to medicine. He set up practice in Boston. In 1843 he established that puerperal fever is contagious and carried from patient to patient by doctors and hospital attendants. This was contested at the time but confirmed later by I.P. Semmelweiss (July 1). Holmes also achieved a literary reputation with his essays, novels and poems.

---

# August 30

*Ernest Rutherford (Nuclear Physics)*
*Birthday: 1871, Nelson, New Zealand*

Ernest Rutherford went from Canterbury College, Christchurch where he worked on the electrical magnetization of iron, to Cambridge, England. There, studying radioactivity of uranium, he found two kinds of radiation, alpha and beta. This led him to his nuclear theory of atomic structure. Rutherford was the founder of nuclear physics and won the 1908 Nobel Prize in Chemistry for his investigations into the disintegration of the elements and the chemistry of radioactive substances.

*Edward Mills Purcell (Physics)*
*Birthday: 1912, Taylorvi e, Illinois*

Edward Mills Purcell studied electrical engineering at Purdue and spent most of his career at Harvard where he earned his doctorate in physics in 1938. He shared the 1952 Nobel Prize in Physics with Felix Bloch for the development of Nuclear Magnetic Resonance (NMR). His research in nuclear magnetic resonance had significant applications in chemical analysis, medical diagnosis, and radio astronomy.

# August 31

*Michel Eugene Chevreul (Chemistry)*
*Birthday: 1786, Angers, France*

Michel Eugene Chevreul received his training in chemistry from Vauquelin and Fourcroy. He lived to be 103 and spent most of his working life in the Museum of Natural History in Paris. He made his most significant contributions through his studies of oils, fats and soaps, analyzing them to discover their chemical composition.

*Herman Helmholtz (Physiology)*
*Birthday: 1821, Potsdam, Germany*

Herman Helmholtz studied medicine at Berlin and became especially interested in physiology. He investigated nerve fibers, nerve cells and nervous impulses and made significant contributions to knowledge about the structure and mechanism of the human eye. His study of animal heat led to investigations into conservation of energy. Interested in sound, he worked on the function of the bones in the middle ear. He was a versatile scientist and incorporated both physics and physiology in his investigations. He made important contributions to the science of theoretical physics.

*Human Eye*

*"The brightest flashes in the world of thought are incomplete until they have been proved to have their counterparts in the world of fact."*

*John Tyndall (English physicist) August 2*

# September

*"Never try to cover up the gaps in your knowledge, even by
the boldest guesses and hypotheses. No matter how this bubble may delight
the eye by its profusion of colors, it is bound to burst,
and you will be left with nothing but confusion."*

*Ivan Petrovich Pavlov (Russian physiologist) September 26*

*Banded Butterflyfish*

# Introduction to September

## Each Day is a New Discovery

My earliest memories of enjoying biology were exploring the creatures that live in the Atlantic ocean where my family used to vacation every summer. One day when I was scooping up sand and watching small sand crabs scurry around, I asked my father what I would have to study to learn about these. He told me "microbiology." When I actually took a microbiology course in college I found out that it wasn't anything like studying the invertebrates of the intertidal and, in fact, I didn't like the course at all. But I found many of my other biology classes to be wonderful, including physiology, comparative anatomy and embryology. When I entered graduate school I only knew that I liked biology but had not yet focused on a particular area of study, nor what career path I would follow. As part of the program we did lab rotations, and one day in the weekly lab meeting of John Trinkaus, I watched a time-lapse movie of cells moving around in culture. I couldn't quite believe what I was seeing and I was hooked! Since that time I have worked on a number of different model systems, but always the fundamental questions concern how cells find their correct position in the embryo. To this day I still take great pleasure in sitting down at the microscope and just watching an embryo develop before my eyes. I still can't quite believe what I am seeing.

I feel every day when I arrive at the lab that I am the luckiest person to have a "job" that I really love. Each day is a new discovery, whether it is a recent experimental result that tells me a little more about how an embryo emerges from a single cell, or a journal article that I have read that challenges my notions about a developmental process. I am fortunate, too, that my position at the University involves teaching at both the undergraduate and graduate level. Preparing for those classes is as much of a learning experience for me as the students, and it is a wonderful "excuse" to share my excitement about cell and developmental biology with others. My career is never boring. It is endlessly fascinating.

Carol Erickson
*Section of Molecular and Cellular Biology*
*University of California-Davis*
*September 2000*

# September

| | | | |
|---|---|---|---|
| 1 | Francis William Aston, 1877 - 1945 | 16 | A. von Szent-Gyorgyi, 1893 - 1986 |
| 2 | James Marsh, 1794 - 1846 | 17 | Stephen Hales, 1677 - 1761 |
| 3 | Matthew Boulton, 1728 - 1809 | 18 | Leon Foucault, 1819 - 1868 |
| | Dixie Lee Ray, 1914 - 1994 | 19 | Jean Baptiste J. Delambre, 1749 - 1822 |
| 4 | Stanford Moore, 1913 - 1982 | 20 | James Dewar, 1842 - 1923 |
| 5 | John Dalton, 1766 - 1844 | 21 | Donald Arthur Glaser, 1926 - |
| 6 | Johan Carl Wilcke, 1732 - 1796 | 22 | Michael Faraday, 1791 - 1867 |
| | Louis Essen, 1908 - 1997 | 23 | Valdemar Poulsen, 1869 - 1942 |
| 7 | William Friese-Greene, 1855 - 1921 | 24 | Robert John Kane, 1809 - 1890 |
| 8 | Derek Harold Richard Barton, 1918 - | 25 | Thomas Hunt Morgan, 1866 - 1945 |
| 9 | Luigi Galvani, 1737 - 1798 | 26 | Ivan Petrovich Pavlov, 1849 - 1936 |
| 10 | Thomas Sydenham, 1624 - 1689 | 27 | Martin Ryle, 1918 - 1984 |
| 11 | Arthur Young, 1741 - 1820 | 28 | Richard Bright, 1789 - 1858 |
| 12 | Irene Joliot-Curie, 1897 - 1956 | | Seymour Cray, 1925 - 1996 |
| 13 | Walter Reed, 1851 - 1902 | 29 | Enrico Fermi, 1901 - 1954 |
| 14 | John Gould, 1804 - 1881 | 30 | Hans Wilhelm Geiger, 1882 - 1945 |
| 15 | Neil Bartlett, 1932 - | | |

# September 1

*Francis William Aston (Chemistry)*
*Birthday: 1877, Harborne, Birmingham, England*

Francis William Aston studied chemistry at Birmingham and worked there for two years on optical rotations. In 1910 he joined J.J. Thomson (December 18) at the Cavendish in Cambridge returning there after the war to develop the mass spectrograph. This proved an invaluable tool in the rapidly advancing nuclear physics research. Aston was awarded the 1922 Nobel Prize in Chemistry for his discovery, by means of his mass spectrograph, of isotopes in a large number of nonradioactive elements and for his enunciation of the whole-number rule.

# September 2

*James Marsh (Chemistry)*
*Birthday: 1794, Birthplace not known, England*

James Marsh was a chemist at the Royal Arsenal, Woolwich, London where he worked as an assistant to Michael Faraday (September 22). His name is known in connection with studies on poisons. In 1836 he developed a test for the presence of arsenic and the test was used forensically with the result that he was made famous by his involvement in murder trials.

# September 3

*Matthew Boulton (Engineering, Manufacturing)*
*Birthday: 1728, Birmingham, England*

Matthew Boulton left school at 14 to join his father's business. His interest in science was kindled by Benjamin Franklin (January 17) who visited Birmingham in 1758, and by meeting Erasmus Darwin (December 12). In 1768 he met James Watt (January 19) hoping to take over the development of Watt's steam engine from James Roebuck. This took some years but eventually he produced a reliable, economical engine to Watt's design.

## September 3

*Dixie Lee Ray (Marine Biology)*
*Birthday: 1914, Tacoma, Washington, USA*

Dixie Lee Ray received her doctorate in biology from Stanford University in 1945 and went to work in the Zoology Department at the University of Washington. Her research focused on comparing marine shore life in various areas of the world. In addition to being a successful marine biologist Dixie Lee Ray also became Chair of the National Atomic Energy Commission, Assistant Secretary of State, and Governor of the State of Washington.

*Banded butterflyfish*

## September 4

*Stanford Moore (Biochemistry)*
*Birthday: 1913, Chicago, Illinois, USA*

Stanford Moore studied chemistry at Vanderbilt and Wisconsin and did research studying proteins at the Rockefeller Institute. With W.H. Stein he developed a method of chromatographic fractionating of hydrolyzed proteins to identify and sequence their constituent amino acids. In 1959 he sequenced the 124 amino acids of ribonuclease. He shared with Stein the 1972 Nobel Prize in Chemistry for their contribution to the understanding of the connection between chemical structure and catalytic activity of the active center of the ribonuclease molecule.

## September 5

*John Dalton (Chemistry)*
*Birthday: 1766, Eaglesfield, Cumberland, England*

John Dalton had little formal schooling but had the good fortune to associate with some of the leading scientists of the day. His great contribution to science was his atomic theory in which he stated that atoms of different elements are distinguished by differences in their weights. He developed his theory in lectures and in his *New System of Chemical Philosophy* (1808).

# September 6

## Johan Carl Wilcke (Physics)
### Birthday: 1732, Wismar, Mecklenburg, Germany

Johan Carl Wilcke grew up in Sweden and studied at Uppsala and Berlin. He did some important work in electricity but is best known for his work on heat. In 1772 he published a paper about the drop in temperature of snow as it melts. This led to the discovery of latent and specific heat. Joseph Black (1728-1799) and Henry Cavendish (October 10) were working in the same area but Wilcke was the first to publish his findings.

*Melting snow*

## Louis Essen (Physics)
### Birthday: 1908, Nottingham, England

Louis Essen studied at the University of London and worked at the National Physical Laboratory in Teddington. His work focused on the physics of frequency generation and measurement. He developed the Essen quartz ring clock in 1938 and in 1955 developed the revolutionary cesium-beam atomic clock.

# September 7

## William Friese-Greene (Invention)
### Birthday: 1855, Bristol, England

William Friese-Greene, a portrait photographer, also made lantern slides for lantern shows. He became interested in the idea of moving pictures and in collaboration with engineer Mortimer Evans in 1889 patented a motion picture camera which took 10 pictures per second on a roll of sensitized paper. He was a pioneer but his apparatus was inferior to that of the Lumiere brothers (October 19).

# September 8

*Derek Harold Richard Barton (Chemistry)*
*Birthday: 1918, Gravesend, Kent, England*

Derek Harold Richard Barton received his degree at London University and became professor of organic chemistry at Burbeck College. He worked on the three-dimensional structures of organic molecules and discovered a method of synthesizing the hormone aldosterone (1960) and shared the 1969 Nobel Prize in Chemistry for his contribution to the development of the concept of conformation and its application in chemistry.

# September 9

*Luigi Galvani (Biology)*
*Birthday: 1737, Bologna, Italy*

Luigi Galvani was an anatomist and was appointed to a chair at Bologna. He is famous for his discovery of animal electricity as a result of the chance observation of the twitching of a frog laid out for dissection on a table that also held an electrical device. Following up on this unexplained movement, Galvani deduced - wrongly as was shown by Volta (February 18) - that animal bodies generate actual electrical currents.

# September 10

*Thomas Sydenham (Medicine)*
*Baptism: 1624, Wynford Eagle, Dorset, England*

Thomas Sydenham, a physician in Westminster, London, is said to have professed little faith in anatomy, pathology or microscopes. Instead he relied upon accurate observations and recorded them very precisely. He used the records to develop clinical descriptions of diseases. He pioneered the use of quinine as a treatment for malaria, and wrote an important paper on gout. He has been called the English Hippocrates because he re-introduced the observational methods practiced by the Greek Hippocrates in the 5th century BC.

# September 11

*Arthur Young (Agriculture)*
*Birthday: 1741, London, England*

Arthur Young was a farmer and wrote extensively about farming in various parts of England, Ireland and France. In 1784 he started the Annals of Agriculture continuing with that publication until 1815. Young's importance is in his contribution to increasing knowledge and understanding of agricultural science. He was made Secretary to the Board of Agriculture in 1793.

# September 12

*Irene Joliot-Curie (Radiochemistry)*
*Birthday: 1897, Paris, France*

Irene Joliot-Curie was the daughter of Marie and Pierre Curie. With her husband Frederick Joliot, she was awarded the 1935 Nobel Prize in Chemistry for their discovery of artificial radioactivity by the production in the laboratory of unstable isotopes of phosphorus. Irene died at age 59 of leukemia caused by prolonged exposure to radiation.

*Mosquito*

# September 13

*Walter Reed (Medicine)*
*Birthday: 1851, Belroi, Virginia, USA*

Walter Reed studied medicine at Virginia and in New York and then joined the US Medical Corps. He did significant studies on erysipelas, diphtheria, and typhoid, but his greatest contribution was made when stationed in Havana. Here he established that yellow fever is transmitted by mosquitoes and that the causative organism was non-filterable (later identified as a virus.)

## September 14

*John Gould (Ornithology)*
*Birthday: 1804, Lyme Regis, Dorset, England*

John Gould learned bird-stuffing as a teenager and in 1827 was made taxidermist to the London Zoological Society. His wife was a talented illustrator and together they produced many beautifully illustrated folios and monographs including *Birds of Europe, Birds of Australia, Mammals of Australia,* a book on humming birds, and many others.

## September 15

*Neil Bartlett (Chemistry)*
*Birthday: 1932, Newcastle-upon-Tyne, England*

Neil Bartlett graduated in chemistry from Durham University and studied at British Columbia, Princeton, then UC Berkeley. His research centered on a family of gasses known as the noble gasses found in the atmosphere in the 1890s by William Ramsay and believed to be totally inert. In 1962 Bartlett prepared a compound of xenon, one of the noble gasses but clearly not inert.

## September 16

*Albert von Szent-Gyorgyi (Biochemistry)*
*Birthday: 1893, Budapest, Hungary*

Albert von Szent-Gyorgyi studied medicine at Budapest and did post graduate studies under Hopkins at Cambridge and Kendall at Rochester. He emigrated to the US in 1947. For identifying ascorbic acid as being identical to vitamin C and explaining its role in respiration, he was awarded the 1937 Nobel Prize in Physiology or Medicine for his discoveries in connection with biological combustion processes, with special reference to vitamin C and the catalysis of fumaric acid.

*"Discovery consists of seeing what everybody has seen*
*and thinking what nobody has thought."*

*Albert Szent-Gyorgyi (U.S. biochemist) September 16*

## September 17

*Stephen Hales (Biology)*
*Birthday: 1677, Beksbourne, Kent, England*

Stephen Hales studied at Cambridge shortly after Isaac Newton (December 25) left there. He was interested in animal and plant respiration and did experiments on plant absorption of water, and the part played by light in plant growth. Hales is popularly remembered as being the first to measure blood pressure. He did so by inserting a glass tube into a vein or artery and noting how high the blood rose in the tube.

## September 18

*Leon Foucault (Physics)*
*Birthday: 1819, Paris, France*

Leon Foucault became physicist at the Paris Observatory in 1855 after publishing (1850) an account of his pendulum experiment to measure the Earth's rotation. His calculation for this experiment became known as Foucault's Law and others repeated his experiment with longer pendulums. He also measured the speed of light using rotating mirrors in a confined situation.

## September 19

*Jean Baptiste Joseph Delambre (Mathematics)*
*Birthday: 1749, Amiens, France*

Jean Baptiste Joseph Delambre received a literary education and then turned to mathematics and astronomy. In his lifetime he was best known for assisting in the measurement of the meridian arc between Barcelona and Dunkirk. However, his major contribution to science was a meticulously detailed series of histories of astronomy.

# September 20

*James Dewar (Chemistry)*
*Birthday: 1842, Kincardine-on-Forth, Scotland*

James Dewar studied at Edinburgh under J.D. Forbes and Lyon Playfair (May 21). In about 1772 he invented the vacuum flask and in following years worked on trying to obtain lower and lower temperatures, in particular he tried to liquefy hydrogen. In 1895 he managed to cool hydrogen to -200 degrees C and saw some liquid hydrogen but did not manage to collect any until 1898.

# September 21

*Donald Arthur Glaser (Physics)*
*Birthday: 1926, Cleveland, Ohio, USA*

Donald Arthur Glaser obtained his Ph.D. at Cal Tech in 1950 studying under Carl Anderson. Moving to Michigan, Glaser began searching for an alternative to a cloud chamber for tracking atomic particles. He received the 1960 Nobel Prize in Physics for his invention of a bubble chamber in which bubbles form along the tracks of particles traveling through a super-heated liquid.

# September 22

*Michael Faraday (Physics)*
*Birthday: 1791, Newington, Surrey, England*

Michael Faraday had no formal education and was apprenticed at 14 to a bookseller. A copy of the Encyclopedia Britannica excited his interest and he began to study science. He made two very important discoveries: electricity is generated when a wire is moved in the field of a magnet; and a magnetic field can cause a ray of polarized light to rotate. His explanations were not readily accepted by his contemporaries.

*Michael Faraday*

## September 23

*Valdemar Poulsen (Invention)*
*Birthday: 1869, Copenhagen, Denmark*

Valdemar Poulsen was a technical advisor to the Copenhagen Telephone Company. He wrote about telephony and telegraphy but is best known for his invention of an audio recording device which recorded sound as variations in the magnetic field along a metal wire. This was an interesting technological advance but the sound quality was inferior to Edison's gramophone.

## September 24

*Robert John Kane (Chemistry)*
*Birthday: 1809, Dublin, Ireland*

Robert John Kane studied pharmacy in Paris and became professor of chemistry in Dublin in 1831. He studied the ethyl radical and synthesized mesitylene. He published the major text of the day, *Elements of Chemistry* (1840) and later, *The Industrial Resources of Ireland*. This later book attracted the attention of politician Sir Robert Peel and Kane was made adviser to the Irish Government.

## September 25

*Thomas Hunt Morgan (Genetics)*
*Birthday: 1866, Lexington, Kentucky, USA*

Thomas Hunt Morgan studied biology at Kentucky and Johns Hopkins Hospital. He was skeptical of Mendel's Laws of Heredity and tested their validity by hybridizing generations of the fruit fly *Drosophila*. His experiments validated Mendel's Laws and in the process Morgan found that some traits were sex linked. He suggested the X and Y chromosome pairing theory and won a 1933 Nobel Prize in Physiology and Medicine for his discoveries concerning the role played by the chromosome in heredity.

*X and Y Chromosomes*

# September 26

*Ivan Petrovich Pavlov (Physiology)*
*Birthday: 1849, Ryazan, Central Russia*

Ivan Petrovich Pavlov studied medicine at St Petersburg and was made Director of Physiology there. He did research in two areas. In his study of digestion, he described three phases of digestion, nervous, pyloric, intestinal. For this work he won a 1904 Nobel Prize in Physiology or Medicine in recognition of his work on the physiology of digestion, through which knowledge on vital aspects of the subject has been transformed and enlarged. His other research for which he is popularly better known, was on the conditioned (learned) reflex.

# September 27

*Martin Ryle (Physics)*
*Birthday: 1918, Brighton, Sussex, England*

Martin Ryle was educated at Oxford. He did war time (WW2) radar research which resulted in successful jamming of V2 rocket guidance systems. After the war he went to the Cavendish in Cambridge and founded the school of radioastronomy. He developed interferometer techniques and produced catalogs of cosmic radio sources. He shared the 1974 Nobel Prize in Physics for his observations and inventions, in particular of the aperture synthesis technique.

# September 28

*Richard Bright (Medicine)*
*Birthday: 1789, Bristol, England*

Richard Bright studied medicine at Edinburgh, traveled to Iceland and the Continent, and then settled into a career at Guy's Hospital in London. He was physician to Queen Victoria in 1837. Bright was a founder of the biochemical study of disease. He described and named Bright's disease (chronic nephritis) and several intestinal and brain disorders.

# September 28

*Seymour Cray (Electronics)*
*Birthday: 1925, Chippewa Falls, Wisconsin, USA*

Considered the father of the supercomputer industry, Cray graduated from the University of Minnesota, worked first for Engineering Research Associates (ERA), then helped found Control Data Corporation (CDC), and later, Cray Research. The first supercomputer, the Cray-1 built in 1976, was 10 times faster than any other computer of the time.

# September 29

*Enrico Fermi (Nuclear Physics)*
*Birthday: 1901, Rome, Italy*

Enrico Fermi received his doctorate at Piza, worked with Max Born (December 11) in Germany, and was made professor of theoretical physics in Rome. He showed that radio active isotopes of most elements could be produced by neutron bombardment. He won the 1938 Nobel Prize in Physics for his demonstrations of the existence of new radioactive elements produced by neutron irradiation, and for his related discovery of nuclear reactions brought about by slow neutrons. In 1942, in the US, he built the first atomic pile - a controlled, self-sustaining nuclear reaction.

# September 30

*Hans Wilhelm Geiger (Nuclear Physics)*
*Birthday: 1882, Neustadt, Germany*

Hans William Geiger received his doctorate at Erlanger studying electrical ionization of gasses and radioactive disintegration. In 1908, with Ernest Rutherford (January 5), he devised a counter for alpha particles. In 1928, at Kiel, he improved the sensitivity of the counter producing a device essentially the same as the Geiger Counter that is used today to measure radiation.

*"A scientist can discover a new star but he cannot make one.*
*He would have to ask an engineer to do it for him."*

*Gordon L. Glegg (American engineer) 1969-*

# October

"When it comes to atoms, language can be used only as in poetry.
The poet, too, is not nearly so concerned with
describing facts as with creating images."

Niels Henrik David Bohr (Danish physicist) October 7

*Early light bulb*

# Introduction to October

## Basic Research Solves Environmental Problems

As a small child, I was always trying to catch frogs and tadpoles or just watching things float (or sink) while wading in small backyard fish ponds where I lived in Los Angeles. The fascination and connections to aquatic life became especially firm after my older sister's boyfriend brought over a microscope and opened my eyes to another universe in those ponds. How could I know at the time that the exotic (and invasive) water hyacinths gracing the ponds, and other aquatic weeds, would play a major role in my professional scientific life.

For me, from the earliest age, the amazing nature of water and all the life that it surrounds and supports, made me wonder at how all this works. On entering college, there was never any doubt about biology as my center of interest. Invertebrate zoology and studies of marine algae eventually led to a few graduate courses focused on new concerns with pollution from oil spills and pesticide contamination, a post "Silent Spring" reaction.

My early post-graduate experience at the EPA in Washington DC also opened my eyes to another form of "biological pollution". I became intrigued with abilities of many aquatic plants from other continents to invade our ecosystems and disrupt the natural balance among plants and animals and, of course, all those human needs for adequate, high-quality water. I found here a clear purpose for combining basic research with finding solutions to these aquatic weeds— applied research.

Doing basic research I discovered that the ability of aquatic plants to form distinctly different leaves ("heterophylly"), which allows them to adapt to changing environmental conditions, is controlled by the growth regulator abscisic acid (ABA). This was a novel role for a plant "hormone" thought to be primarily involved with dormancy and senescence. The Agricultural Research Service has afforded an excellent opportunity for such scientific discovery with the added, and for me very important, bonus of applying such discovery to trying to solve an environmental problem.

The spread of exotic plants continues, though there have been real successes through both biological control and use of safe herbicides. But now it's time for much more public awareness, education and responsibility so that the environmental stewardship ethic that has reduced chemical pollutants can be applied to the spread of living, exotic invasive species.

*Lars W. J. Anderson*
*USDA Aquatic Weed Research*
*University of California, Davis*

# October

| | | | |
|---|---|---|---|
| 1 | Otto Robert Frisch, 1904 - 1979 | 17 | Paul Bert, 1833 - 1886 |
| 2 | Ferdinand Gustav J. von Sachs, 1832 - 1897 | | Mae Carol Jemison, 1956 - |
| 3 | Patrick Manson, 1844 - 1922 | 18 | Christian F. Schonbein, 1799 - 1868 |
| 4 | Alice Stewart, 1906 - | 19 | Auguste Lumiere, 1862 - 1954 |
| | Sputnik I, 1957 - | 20 | Christopher Wren, 1632 - 1723 |
| 5 | Robert Hutchings Goddard, 1882 - 1945 | | James Chadwick, 1891 - 1974 |
| 6 | Francoise Magendie, 1783 - 1855 | 21 | Alfred Bernhard Nobel, 1833 - 1896 |
| 7 | Niels Henrik David Bohr, 1885 - 1962 | 22 | Stephen Moulton Babcock, 1843 - 1931 |
| 8 | Otto Heinrich Warburg, 1883 - 1970 | 23 | Felix Bloch, 1905 - 1983 |
| 9 | Pierre Joseph Macquer, 1718 - 1784 | 24 | Antony van Leeuwenhoek, 1632 - 1723 |
| | Herman Emil Fischer, 1852 - 1919 | 25 | Robert Stirling, 1790 - 1878 |
| 10 | Henry Cavendish, 1731 - 1810 | | Richard Evelyn Byrd, 1888 - 1957 |
| 11 | Friedrich Bergius, 1884 - 1949 | 26 | Giovanni Maria Lancisi, 1654 - 1720 |
| 12 | Arthur Harden, 1865 - 1940 | 27 | Isaac Merrit Singer, 1811 - 1875 |
| 13 | Rudolph Ludwig C. Virchow, 1821 - 1902 | 28 | Jonas Edward Salk, 1914 - 1955 |
| 14 | Friedrich Wilhelm Kohlrausch, 1840 - 1910 | | William Gates, 1955 - |
| 15 | Evangelista Torricelli, 1608 - 1647 | 29 | Edmund Halley, 1656 - 1742 |
| 16 | Robert Stephenson, 1803 - 1859 | 30 | Hermann F. Moritz Kopp, 1817 - 1892 |
| | | 31 | Joseph Wilson Swan, 1828 - 1917 |

# October 1

*Otto Robert Frisch (Physics)*
*Birthday: 1904, Vienna, Austria*

Otto Robert Frisch was educated in Vienna. While visiting his aunt, Lise Meitner in 1938, he discussed the results that Otto Hahn (March 8) had achieved with slow neutron bombardment of uranium. Their explanation was that nuclear fission had occurred. Frisch performed experiments that successfully demonstrated the existence of fission fragments. He continued to work on atomic energy and participated with the atom bomb team at Los Alamos.

# October 2

*Ferdinand Gustav Julius von Sachs (Botany)*
*Birthday: 1832, Wroclaw, Silesia*

Ferdinand Gustav Sachs studied at Prague and eventually became professor of botany at Wurzburg. He amassed a large body of work on plant metabolism, much of it being done using water culture. He also investigated phototropism and geotropism. His best know publication is *Lehrbuch der Botanik* (1868) a textbook that describes many of his original investigations.

*Parasitic worm egg*

# October 3

*Patrick Manson (Medicine)*
*Birthday: 1844, Olmeldrum, Scotland*

Patrick Manson was appointed Medical Officer for Formosa and remained in the Far East for 23 years. He introduced vaccination to the region, studied filarial infection finding it to be transmitted by mosquitoes, and studied ringworm and guinea worm. He founded the London School of Tropical Medicine. His suggestion that malaria is spread by mosquito stimulated Ronald Ross (May 13) to investigate this.

# October 4

*Alice Stewart (Medicine)*
*Birthday: 1906, England*

Alice Stewart earned her MD from Cambridge University in 1934. In the early years of World War II she joined the Nuffield Department of Clinical Medicine in Oxford and developed ways of protecting the health of workers in several war-related industries. After the war she set about finding out why unusually large numbers of babies were dying of leukemia. Her unpopular finding issued in 1956 was that the leukemia was associated with a favorite tool of the medical profession - the X-ray. Follow-up studies using larger sample populations confirmed her finding that X-raying pregnant women can lead to leukemia in their children.

*Sputnik I*
*Event: 1957, Space Exploration*

The satellite Sputnik I, launched by the USSR on October 4, 1957, was the first man-made object to orbit Earth. It remained in orbit until January 4, 1958. The satellite was designed to send radio signals to Earth about the density of the upper atmosphere but it only transmitted signals for a short time after launch.

# October 5

*Robert Hutchings Goddard (Physics)*
*Birthday: 1882, Worcester, Massachusetts, USA*

Robert Hutchings Goddard received his Ph.D. at Clark. He was interested in rocketry and in 1926 sent up his first rocket. In 1929 he launched a larger one carrying a barometer, a thermometer, and a camera. From 1930-35 he launched rockets that achieved 550 mph and heights of $1 1/2$ miles. He invented a multi-stage rocket and has been called the father of the space age.

# October 6

*Francoise Magendie (Medicine)*
*Birthday: 1783, Bordeaux, France*

Francoise Magendie studied medicine at Paris and pioneered experimental physiology. He showed that the anterior nerve roots of the spinal cord carry motor messages to muscles, and that posterior roots carry sensory messages to the brain. He showed that life could not be sustained without dietary protein. From 1843 he worked with his assistant Claude Bernard (July 12) on animal heat and blood sugar.

# October 7

*Niels Henrik David Bohr (Physics)*
*Birthday: 1885, Copenhagen, Denmark*

Niels Henrik David Bohr received his doctorate at Copenhagen and worked for several years with Ernest Rutherford (August 30) at Manchester. He received the 1922 Noel Prize in Physics for his explanation of the quantum theory of the electronic structure of the hydrogen atom, and the origin of hydrogen and helium spectral lines. In later years he worked for peaceful uses of atomic energy.

# October 8

*Otto Heinrich Warburg (Biochemistry)*
*Birthday: 1883, Freiburg-im-Breisgan, Germany*

Otto Heinrich Warburg studied chemistry under Emil Fischer (October 9), then studied medicine, and concentrated his research on tissue respiration. He was awarded the 1931 Nobel Prize in Physiology or Medicine for his new insights into the involvement of cytochromes in respiration and for the discovery of ferredoxin, a key factor in photosynthesis. He also investigated metabolism in cancer cells.

*Cancer cell apoptosis*

# October 9

*Pierre Joseph Macquer (Chemistry)*
*Birthday: 1718, Paris, France*

Pierre Joseph Macquer studied medicine in Paris. His contribution was to advance the subject of chemistry more by organizing and imparting information than by great discoveries. However, he studied gypsum, Prussian blue, fermentation, and dyeing, and contributed to the development of the process of making the renowned Sevres porcelain.

*Herman Emil Fischer (Chemistry)*
*Birthday: 1852, Bonn, Germany*

Herman Emil Fischer studied chemistry at Bonn and, in 1882, as professor at Erlangen he worked on the constitution and synthesis of natural products. He is recognized as having been one of the great organic chemists. Early in his career he discovered phenyhydrazine which provided insights into the carbohydrate chemistry. At Erlangen he synthesized purine, d-glucose, and some other sugars, and was an advisor to his government on synthetic foods. He was awarded the 1902 Nobel Prize in Chemistry for his work on sugar and purine synthesis.

# October 10

*Henry Cavendish (Physics)*
*Birthday: 1731, Nice, France*

Henry Cavendish, nephew of the 3rd Duke of Devonshire, attended Cambridge but took no degree. His experiments in electricity anticipated much of what would be discovered in the next 50 years. He demonstrated the existence of hydrogen in the atmosphere, and determined the density of Earth. He is commemorated by the renowned Cavendish Laboratory in Cambridge.

# October 11

*Friedrich Bergius (Chemistry)*
*Birthday: 1884, Goldschmieden, Poland*

Friedrich Bergius studied in Germany under the chemists Nernst (June 25) and Haber and became an outstanding industrial chemist. He was interested in chemical reactions at high pressures and temperatures and developed the Bergius Process of hydrogenation of coal to petroleum for which he shared the 1931 Nobel Prize in Chemistry. He also developed a process to produce alcohol and sugar from wood fibers.

# October 12

*Arthur Harden (Biochemistry)*
*Birthday: 1865, Manchester, England*

Arthur Harden studied at Manchester and Erlangen. With H.E. Roscoe he published *A New View of the Genesis of Dalton's Atomic Theory*. He is notable for his work on the alcoholic fermentation of sugars, for showing that zymase is a complex of enzymes, and for describing the conversion of glycogen to lactic acid in muscle. He shared the 1929 Nobel Prize in Chemistry for investigations on the fermentation of sugar and fermentative enzymes.

# October 13

*Rudolph Ludwig Carl Virchow (Biology)*
*Birthday: 1821, Schivelbein, Germany*

Rudolph Ludvig Carl Virchow studied medicine at Berlin. In 1854 he wrote; "There is no life but through direct succession" (omnis cellula e cellula). He is known as the founder of cellular pathology. He studied tumors and animal parasites of man, and was antagonistic to the theory of the role of bacteria in disease. He was interested in anthropology and archeology and published in both areas.

*Rudolph Ludwig Carl Virchow*

# October 14

*Friedrich Wilhelm Kohlrausch (Physics)*
*Birthday: 1840, Rinteln-on-Weser, Germany*

Friedrich Wilhelm Kohlrausch was educated at Erlangen and Gottingen. He is best known for research on electrical conductivity of solutions at different concentrations using an alternating current. He showed that conductivity increases as concentration decreases. He formulated the Kohlrausch Law of independent migration of ions.

# October 15

*Evangelista Torricelli (Mathematics, Physics)*
*Birthday: 1608, Faenza, Italy*

Evangelista Torricelli was a mathematician and physical scientist. He verified Galileo's laws of falling bodies, became Galileo's assistant just before he died, and succeeded him as mathematician to the court of Tuscany. He ground telescope lenses, constructed a simple microscope, and invented (1643) a mercury barometer. In pure mathematics he worked on conic sections and the cycloid.

# October 16

*Robert Stephenson (Engineering)*
*Birthday: 1803, Willington Quay, England*

Robert Stephenson, son of railway engineer George Stephenson (June 9, 1781), studied for a year at Edinburgh and then joined his father at Stephenson and Company in Newcastle. He built several steam engines including the Lancashire Witch, and designed wrought iron railway bridges in a unique tubular design. His Britannia Bridge over the Menai Straits was a remarkable achievement.

# October 17

*Paul Bert (Biology)*
*Birthday: 1833, Auxerre, France*

Paul Bert qualified in law but turned to science and became assistant to Claude Bernard (July 12) studying the physiology of respiration at high and low atmospheric pressures both in animals and himself. He showed that oxygen at high partial pressures is a poison. Unfortunately, his work although significant, was ignored for about 60 years.

*Mae Carol Jemison (Medicine)*
*Birthday: 1956, Decatur, Alabama, USA*

Mae Carol Jemison studied at Stanford and Cornell and received her MD in 1981. She practiced medicine, joined the Peace Corps, and then deciding to pursue a childhood dream, applied to NASA's astronaut training program. Becoming the first black woman in space, she flew the eight day space shuttle Endeavor STS-47 mission—launch date, September 12, 1992.

# October 18

*Christian Friedrich Schonbein (Chemistry)*
*Birthday: 1799, Metzingen, Germany*

Christian Friedrich Schonbein did not have a university education but studied and attended lectures by such scientists as Faraday (September 22) and Gay-Lussac (December 6) and was elected to the Chair of Chemistry at Basle. He did classic work on ozone, including naming the gas. He prepared cellulose nitrate (gun cotton) and investigated its explosive properties but found no safe way to manufacture it.

## October 19

*Auguste Lumiere (Invention)*
*Birthday: 1862, Besancon, France*

Auguste Lumiere, with his brother Louis, built (1895) the first satisfactory cine-camera and projector. They manufactured 35mm perforated film and a projector with an intermittent claw movement. In 1900 they introduced Photorama which projected a 360 degree picture, in 1904 they patented Autochrome color plates, and in 1935 they introduced stereo projection.

*35mm film*

## October 20

*Christopher Wren (Architecture, Mathematics)*
*Birthday: 1632, East Knoyle, England*

Christopher Wren studied mathematics and astronomy at Wadham College, Oxford and was appointed to the Savilian Chair of Astronomy at Oxford in 1661. He made some minor contributions to mathematics and astronomy until 1665 when he was appointed a commissioner for the restoration of the old St. Paul's Cathedral in London. From that time he gave more and more time to architecture and in all, designed eighty London buildings including the Greenwich Observatory.

*James Chadwick (Physics)*
*Birthday: 1891, Bollington, Cheshire, England*

James Chadwick graduated from Manchester Honors School of Physics in 1911 and worked under Ernest Rutherford (August 30) on radioactivity. He was awarded the 1935 Nobel Prize in Physics for proposing the existence of a neutron particle, verifying its existence, and calculating its mass. In 1943-5 he headed a delegation of British experts working in New Mexico in the U.S. with the team developing the atom bomb.

## October 21

*Alfred Bernhard Nobel (Invention)*
*Birthday: 1833, Stockholm, Sweden*

Alfred Bernhard Nobel, the inventor of dynamite bequeathed his fortune to establish the annual Nobel Prizes. Following the dictates of his will, annual awards are given for peace, literature, physics, chemistry, and physiology or medicine. In 1968 an economics prize was added. These awards are international, financially significant, and most prestigious.

## October 22

*Stephen Moulton Babcock (Chemistry)*
*Birthday: 1843, Bridgewater, New York, USA*

Stephen Moulton Babcock was educated at Tufts and Gottingen. He became chemist to the New York Agricultural Experiment Station and then professor of agricultural chemistry at Wisconsin. He devised a method of estimating the fat content of milk, and worked on pasteurization and curing of cheese, and the nutritional value of animal foodstuffs.

## October 23

*Felix Bloch (Physics)*
*Birthday: 1905, Zurich, Switzerland*

Felix Block received his Ph.D. from Leipzig in 1928 and moved to Stanford in 1934. He was interested in nuclear magnetism and made the first accurate measurement of the magnetic properties of the neutron in 1939. In WWII he did research on radar and later used this experience to measure nuclear magnetic resonance which gives information about chemical composition and structure.

## October 24

### Antony van Leeuwenhoek (Biology, Invention)
### Birthday: 1632, Delft, Netherlands

Antony van Leeuwenhoek was apprenticed at 16 to a linen-draper and later set himself up in business as a draper. After work each day he ground glass lenses and made hand-held microscopes far superior to any others at the time. He was the first to observe protozoa in pond water, and bacteria in the human mouth. He described himself as having a craving after knowledge more than most men.

*Protozoa*

## October 25

### Robert Stirling (Engineering, Invention)
### Birthday: 1790, Cloag, Scotland

Robert Stirling was educated at Glasgow and Edinburgh, and ordained in 1816. He was interested in mechanics and, with his brother James, invented and patented a unique hot air engine. The Stirling engine was an external combustion engine with a closed piston and cylinder arrangement. Interest in the engine revived in the 1970s and a 1985 French Submarine was equipped with two Stirling engines.

### Richard Evelyn Byrd (Geography)
### Birthday: 1888, Winchester, Virginia, USA

Richard Evelyn Byrd, a navy pilot whose qualifications are administrative and adventurous more than they are scientific, crossed the Atlantic in a dirigible in 1921, took part in a 1925 polar expedition, was the first to fly over the North Pole in 1926, and spent 1928-1955 mapping Antarctica.

# October 26

*Giovanni Maria Lancisi (Medicine)*
*Birthday: 1654, Rome, Italy*

Giovanni Maria Lancisi graduated as a Doctor of Medicine in 1672 (at age 18) from the Sapienza in Rome and in 1696 became professor of medicine there. He laid the foundations of cardiac pathology in two works: *On Sudden Death,* and *On the Motion of the Heart* and *On Aneurysms.* He also wrote on epidemic fevers and suggested that mosquitoes played a part in transmission of malaria.

*Mosquito*

# October 27

*Isaac Merrit Singer (Invention)*
*Birthday: 1811, Pittsdown, New York, USA*

Isaac Merrit Singer was employed as unskilled labor at age 12 and eventually became a traveling mechanist. He took out patents for rock-drilling, and metal- and wood-shaping equipment. He improved upon a sewing machine designed by Elias Howe and took out his own patent for a Singer Sewing Machine in 1851. He continued improving this machine for the next 12 years.

# October 28

*Jonas Edward Salk (Microbiology)*
*Birthday: 1914, New York, USA*

Jonas Edward Salk studied medicine at New York and worked as a microbiologist at the Michigan School of Public health, and Pittsburgh School of Medicine. He developed a method of killing the polio virus in such a way that it would not cause the disease but would cause the production of antibodies against itself. The Salk Poliomyelitis vaccine was first used in quantity in 1954.

## October 28

*William Gates (Computer Science)*
*Birthday: 1955, Seattle, Washington, USA*

William (Bill) Gates, with his friend Paul Allen, was in the computer business before he finished high school. He entered Harvard in 1973 to read law but dropped out in 1975 and, with Allen, created a software computer company called Microsoft. Within eighteen months they had made several hundred thousand dollars. Bill's friends call him brilliant. His company dominates the market and he is now one of the wealthiest men in the world.

## October 29

*Edmund Halley (Astronomy)*
*Birthday: 1656, London, England*

Edmund Halley was educated at Oxford. He did not take a degree but was later awarded an honorary MA by Oxford. His friends included the Astronomer Royal John Flamsteed (August 19), Robert Hooke (July 18), and Isaac Newton (December 25). He made important discoveries about the nature of comets recognizing their periodicity and predicted the 1758 return of the comet bearing his name but did not live to see it.

## October 30

*Hermann Franz Moritz Kopp (Chemistry)*
*Birthday: 1817, Hanau, Germany*

Hermann Franz Moritz Kopp, son of a physician, studied chemistry. His researches focused on the physical and chemical properties of elements and compounds. He is most often recognized as a historian of chemistry and published Gesichte der Chemie (1843-47).

# October 31

*Joseph Wilson Swan, (Invention)*
*Birthday: 1828, Sunderland, County Durham, England*

Joseph Wilson Swan was apprenticed to a chemist and later became partner in a chemical supply company. He is best remembered for his development of improved incandescent filament lamps. He used carbon filaments and then a filament made by extruding nitrocellulose dissolved in acetic acid. Following a patent conflict with Edison (February 11), the Edison and Swan United Electric Light Company was formed.

*Early light bulb*

*"The most beautiful experience we can have is the mysterious. It is the fundamental emotion which stands at the cradle of true art and true science. Whoever does not know it and can no longer wonder, no longer marvel, is as good as dead, and his eyes are dimmed."*

*Albert Einstein (German/American physicist) March 14*

# November

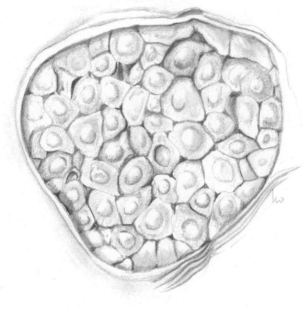

*"I am among those who think that science has great beauty.
A scientist in his laboratory is not only a technician: he is also a child
placed before natural phenomena which impress him like a fairy tale." (1933)*

*Marie Curie (Polish physicist) November 7*

*Islet cells*

# Introduction to November

## Curiosity - the Final Frontier

As I grew up in rural New England, I loved walking through the forest. I knew my area so well that as a pre-teen, a USGS geologist used me as a guide to discover glacial erratics hidden in the woods. But I never connected my love of nature with the fascinating books on Physics and Chemistry my grandfather sent me from New York each year at Christmas - concepts fascinating but so remote from anything I could conceive.

I was going to be a veterinarian until as a senior in High School, I had my first class in Physics. My teacher, a wily old priest, took me aside and helped me become a ham radio operator, supported me in science fairs, and convinced me that I could contribute to science as first revealed to me in my grandfather's books. I developed nuclear and atomic techniques for tracing air pollution, founded a new national visibility network, and spearheaded a major clean up of power plants near Grand Canyon National Park.

What has kept me on the path is a curiosity to go farther into the unknown for questions that matter, sure only that the answer is not at the back of the chapter! And as I went from nuclear astro-physics to applied physics, then atmospheric physics, and now global climate change, I ended up where I began. The quest for knowledge of climate change has joined the urgency of protecting the woods of my youth. I can now, as Robert Frost noted, "...unite my avocation and my vocation.".

*Thomas A. Cahill*
*DELTA Group, Atmospheric Sciences(LAWR)/Physics*
*University of California, Davis*
*October 2000*

# November

| | | | |
|---|---|---|---|
| 1 | Alfred Lothar Wegener, 1880 - 1930 | 14 | Charles Lyell, 1797 - 1875 |
| | | | Frederick G. Banting, 1891 - 1941 |
| 2 | George Boole, 1815 - 1864 | | |
| 3 | Frederick Stratten Russell, 1897 - 1984 | 15 | William Herschel, 1738 - 1822 |
| 3 | Mariner 10, 1973 | 16 | Jean le Rond D'Alembert, 1717 - 1783 |
| 4 | Frederick Orpen Bower, 1855 - 1948 | 17 | August Ferdinand Mobius, 1790 - 1868 |
| 5 | John B. Sanderson Haldane, 1892 - 1964 | 18 | Louis J. Mande Daguerre, 1787 - 1851 |
| 6 | Ian Morris Heilbron, 1886 - 1959 | 19 | George Emil Palade, 1912 - |
| 7 | Marya (Marie) Sklodowska Curie, 1867 - 1934 | 20 | Edwin Powell Hubble, 1889 - 1953 |
| | Lise Meitner, 1878 - 1968 | 21 | Johann August Brinell, 1849 - 1925 |
| 8 | Kate Sessions, 1857 - 1940 | 22 | Louis Eugene Felix Neel, 1904 - |
| | Christian Neethling Barnard, 1922 - | 23 | Henry G. Jeffreys Moseley, 1887 - 1915 |
| 9 | Florence Rena Sabin, 1871 - 1953 | 24 | William C. Williamson, 1816 - 1895 |
| | Carl Edward Sagan, 1934 - 1996 | 25 | Julius Robert von Mayer, 1814 - 1878 |
| 10 | Francis Maitland Balfour, 1851 - 1882 | 26 | Norbert Wiener, 1894 - 1964 |
| 11 | Marie Francois Xavier Bichat, 1771 - 1802 | 27 | Anders Celsius, 1701 - 1744 |
| 12 | Jacques A. Cesar Charles, 1746 - 1823 | 28 | John Wesley Hyatt, 1837 - 1920 |
| 13 | Johanna Gabrielle Edinger, 1897 1967 | 29 | John Ray, 1628 - 1705 |
| | First Artificial Snowstorm, 1946 | 30 | Robert Broom, 1866 - 1951 |

# November 1

*Alfred Lothar Wegener (Geology)*
*Birthday: 1880, Berlin, Germany*

Alfred Lothar Wegener received his doctorate at Berlin, spent two years in Greenland as a meteorologist with a Danish expedition, and returned to Greenland several times after that. He is best remembered for his theory of continental drift (1912) which was controversial for many years but is now widely accepted.

*Pangea's Breakup –*
*Continential Drift*

# November 2

*George Boole (Mathematics)*
*Birthday: 1815, Lincoln, England*

George Boole, son of a small shopkeeper, was largely self-educated. He was appointed to the chair of mathematics at Queen's College, Cork. His two books on mathematics dealt with differential equations and the calculus. He is best known for his contribution of mathematical logic and Boolean algebra - concepts that were very new.

# November 3

*Frederick Stratten Russell (Biology)*
*Birthday: 1897, Bridgport, Dorset, England*

Frederick Russell was educated at Cambridge. After WW I, Russell joined the scientific staff of the Marine Biological Association at Plymouth. He is best known for his work to understand life histories and distribution of plankton. He was particularly interested in several species of medusa and published in two volumes, *The Medusae of the British Isles* (1953, 1970).

# November 3

*Mariner 10*
*Event: 1973, US Space Flight*

Mariner 10, an unmanned probe, was launched by the US on November 3, 1973. It passed Venus on February 4, 1974 and arrived at its destination, Mercury, seven weeks later on March 29, 1974. This was the first time that the gravity of one planet (Venus) had been used to whip a spacecraft to another (Mercury).

# November 4

*Frederick Orpen Bower (Botany)*
*Birthday: 1855, Ripon, Yorkshire, England*

Frederick Orpen Bower studied botany at Cambridge, and in Germany under J. von Sachs. As a plant morphologist he concentrated his interest on Pteridophyta and published *The Ferns* in three volumes (1923-28). His major contribution after he retired from the chair of botany in Glasgow was *Primitive Land Plants* (1935).

# November 5

*John Burdon Sanderson Haldane (Genetics)*
*Birthday: 1892, Oxford, England*

John Burdon Sanderson Haldane (son of the eminent physiologist John Scott Haldane) studied at Oxford and was at University College London for 20 years. His research was in genetics and human physiology and formed a basis for mathematical studies of population genetics. He wrote *Enzymes* (1930), *The Causes of Evolution* (1932), and *The Biochemistry of Genetics* (1954).

# November 6

## Ian Morris Heilbron (Chemistry)
### Birthday: 1886, Glasgow, Scotland

Ian Morris Heilbron's research in organic chemistry in Glasgow was in terpenes found in essential oils and oleo resins in such plants as camphor and eucalyptus. His main contributions were in the natural product field including spectroscopic and chromatographic investigations of vitamins A and D, penicillin, and some dyestuffs. He published a *Dictionary of Organic Compounds*.

# November 7

## Marya (Marie) Sklodowska Curie (Physics, Radiochemistry)
### Birthday: 1867, Warsaw, Poland

Marie Curie was awarded two Nobel prizes; the 1903 Prize in Physics, shared with her husband Pierre, for researches on radiation phenomena; and the 1911 Prize in Chemistry for her discovery of radium and polonium and the isolation of radium and the study of its nature and compounds. She was the first woman professor in France. With her discovery of radium she introduced the scientific world to radioactivity and the fact that the atom is not a solid piece of matter.

## Lise Meitner (Nuclear Physics)
### Birthday: 1878, Vienna, Austria

Lise Meitner was director of a center for radiation physics in Berlin for twenty years. She was Germany's first woman physics professor. At the age of sixty she explained that the uranium atom could split and in doing so, release a tremendous amount of energy. For the discovery that she initiated and explained, her German colleague Otto Hahn was awarded a Nobel prize.

*Lise Meitner*

# November 8

*Kate Sessions (Horticulture)*
*Birthday: 1857, San Francisco, California, USA*

Kate Sessions earned a degree in chemistry from UC Berkeley and then opened a plant nursery in Coronado, California. She traveled throughout the world collecting new plants including the Erythea palm, flame eucalyptus, Chinese twisted juniper, and various acacias. She created Balboa Park in San Diego in 1892.

*Christian Neethling Barnard (Surgery)*
*Birthday: 1922, Beaufort West, South Africa*

Christian Neethling Barnard qualified in 1946 as a physician and joined Groot Schurr Hospital in Cape Town. He developed a way (with cold water circulating through a balloon inserted into the stomach) of lowering body temperature to slow body processes during surgery. He did the first heart transplant in a dog, and then, on December 3, 1967, the first human heart transplant.

# November 9

*Florence Rena Sabin (Medicine)*
*Birthday: 1871, Central City, Colorado, USA*

Florence Sabin graduated from Smith College and then was the first woman to graduate in medicine from Johns Hopkins. She was also the first woman to join the medical faculty there. Sabin studied the lymphatic system, blood cells, and blood vessels, monocytes, and the way the immune system responds to the tuberculosis bacillus. By 1919 she had determined the embryonic origin of red blood corpuscles.

# November 9

*Carl Edward Sagan (Astrophysics)*
*Birthday: 1934, New York City, USA*

Sagan earned four degrees from Chicago University graduating finally with a doctorate in astronomy and astrophysics. He was a prolific writer of popular science and had a great impact upon the general publics of the world through his books and television programs.

# November 10

*Francis Maitland Balfour (Biology)*
*Birthday: 1851, Edinburgh, Scotland*

Francis Maitland Balfour studied at Cambridge under Michael Foster and became interested in the embryological development of Elasmobranch (cartilaginous) fishes. His major published work, the two-volume *Comparative Embryology* (1880, 1881) was very highly regarded and Oxford and Edinburgh each offered him chairs but he preferred to stay at Cambridge.

# November 11

*Marie Francois Xavier Bichat (Biology)*
*Birthday: 1771, Thoirette, Jura, France*

Marie Francois Xavier Bichat studied medicine at Montpellier and Lyons. He lectured and practiced, and conducted autopsies. He made a significant contribution to science in recognizing 21 different body tissues and noting that a single organ might consist of several types of tissue. At this time animal cells had not been recognized and Bichat's work laid the groundwork for the science of histology.

*Capillary vessels*

# November 12

*Jacques Alexandre Cesar Charles (Invention)*
*Birthday: 1746, Beaugency, France*

Jacques Alexandre Cesar Charles was intrigued by Benjamin Franklin's experiments and familiar with the work of Priestly and Cavendish. He experimented with balloons filled with hydrogen sending up a balloon and passenger in 1783 which reached a height of 3000 meters. Charles is also remembered for his inventions of some ingenious physical science instruments.

# November 13

*Johanna Gabrielle Edinger (Paleontologist)*
*Birthday: 1897, Frankfurt, Germany*

Johanna Gabrielle Edinger earned her doctorate in 1921 at the University of Frankfurt and worked there as a research assistant in paleontology. In 1927 Tilly Edinger became curator of the vertebrate collection at the Frankfurt Senckenberg Museum and two years later published Die Fossilen Gehirne (Fossil Brains) in which she virtually invented the field of paleoneurology. Immigrating to the U.S. in 1939 and working at the Harvard Museum of Comparative Zoology, Edinger earned a reputation as an eminent figure in vertebrate paleontology.

*First Artificial Snowstorm (Invention)*
*Event: 1946, Schnectady, New York, USA*

The first artificial snowstorm was produced in 1946 by Vincent Joseph Schaefer. Schaefer worked in a machine shop at General Electric Corporation in Schenectady. In the 1940s, investigating the cause of icing of airplane wings at high altitudes he dropped dry ice into his apparatus and created a miniature snowstorm. On November 13, 1946 he replicated this unexpected result on a much larger scale above the clouds SE of Schenectady.

# November 14

*Charles Lyell (Geology)*
*Birthday: 1797, Kinnordy, Scotland*

Charles Lyell was much influenced at Oxford by geologist William Buckland and, after a brief period of studying law, focused his life's work on geology studies. His special contribution to science was his three-volume *Principles of Geology* (1830, 1832, and 1833). In these writings he stressed the theory that geological history is a matter of ordinary forces and unlimited time (Uniformitarianism) rather than successive world-wide catastrophes which was the prevailing opinion of the time.

*Frederick Grant Banting (Medicine)*
*Birthday: 1891, Alliston, Ontario, Canada*

Frederick Grant Banting qualified as a doctor at Toronto in 1916 and practiced as a surgeon in London. In 1920 he read an article on the connection between the pancreatic Islets of Langerhans and diabetes. With the help of student Charles Best he extracted and named insulin from the Islets. Banting shared the 1923 Nobel Prize for the discovery of insulin with Canadian John Macleod.

*Islet Cells*

# November 15

*William Herschel (Astronomy)*
*Birthday: 1738, Hanover, Germany*

William Herschel made his living as an organist and musician in England and about 1766 he started to studied mathematics and astronomy in his spare time. Using telescopes that he made himself, Herschel discovered, in 1781, the planet Uranus which he believed at first to be a comet. He also discovered the intrinsic movement of the Sun, and infra-red in sunlight.

# November 16

*Jean le Rond D'Alembert (Mathematics)*
*Birthday: 1717, Paris, France*

Jean le Rond d'Alembert spent some time studying law and medicine but eventually turned to mathematics and mechanics. He contributed to calculus, and in applied mathematics he obtained general equations for the motion of fluids, and developed a theory of vibrating strings. He also assisted in the preparation of Diderot's politically controversial encyclopedia (1751-72).

# November 17

*August Ferdinand Mobius (Astronomy)*
*Birthday: 1790, Schulpforta, Germany*

August Ferdinand Mobius first studied law but soon turned to astronomy and mathematics. He became professor of astronomy at Leipzig and director of the observatory. He published Die Elemente der Mechanik der Himmels (1843), a serious work on celestial mechanics. He was also interested in topology and is popularly remembered by the single-sided Mobius Strip.

# November 18

*Louis Jacques Mande Daguerre (Invention)*
*Birthday: 1787, Cormeilles-en-Parisis, France*

Louis Jacques Mande Daguerre was a stage designer and invented the Diorama which incorporated huge paintings on semi-transparent linen. Preliminary sketches on the linen were made with the camera obscura method and Daguerre devised a process of chemical fixation of the images using the fact that light causes silver iodide to darken. His invention was also used for photographic portraits.

# November 19

*George Emil Palade (Biology)*
*Birthday: 1912, Jassy, Moldavia (Iasi, Romania)*

George Emil Palade studied medicine at Bucharest and became a U.S. citizen in 1952. Working at the Rockefeller Institute he used the electron microscope to study organelles in intact cells. By 1956 he had shown that microsomes were rich in RNA and they were renamed ribosomes. For the discovery of two new organelles (lysosomes and peroxisomes) he shared a 1974 Nobel Prize.

# November 20

*Edwin Powell Hubble (Astronomy)*
*Birthday: 1889, Marshfield, Missouri, USA*

Edwin Powell Hubble graduated in law from Chicago and was a Rhodes Scholar at Oxford. He then turned to his avocation, astronomy, at Yerkes and Mount Wilson. He showed that spiral nebulae exist outside our Galaxy, calculated distances for most visible nebulae, and showed that extragalactic nebulae were receding at a predictable rate according to the calculation called Hubble's Law.

# November 21

*Johann August Brinell (Engineering)*
*Birthday: 1849, Bringetofta, Sweden*

Johann August Brinell attended a technical school and while working as a designer and iron-works engineer, he became interested in metallurgy. In 1885 he wrote about textural changes of steel during heating and cooling. He developed an instrument to determine the hardness of steel and experimented on abrasion resistance of various materials.

# November 22

*Louis Eugene Felix Neel (Physics)*
*Birthday: 1904, Lyons, France*

Louis Eugene Felix Neel studied at Strasburgh doing research into the magnetic properties of solid materials. In 1932 he demonstrated a fourth type of magnetism, with a small magnetic effect, that he called antiferromagnetism. In 1948 he explained the strong magnetic effects in what he called ferrimagnetic materials. He shared the 1970 Nobel Prize in Physics.

# November 23

*Henry Gwyn Jeffreys Moseley (Chemistry)*
*Birthday: 1887, Weymouth, Dorset, England*

Henry Gwyn Jeffreys Moseley studied natural science at Oxford. In 1914 he showed, by his investigations of X-ray spectra of over 30 elements, that there were discontinuities in the series suggesting that some elements were missing. Moseley made the connection between nuclear charge and atomic number.

*Fossil plant*

# November 24

*William Crawford Williamson (Biology)*
*Birthday: 1816, Scarborough, England*

William Crawford Williamson had little formal education but was apprenticed to an apothecary, and learned some medicine in Manchester and London. He also acquired a considerable body of knowledge of natural history and investigated invertebrate anatomy, the structure of fish scales, and fossil plants found in the British Coal Measures. He helped to found the science of palaeobotany.

# November 25

*Julius Robert von Mayer (Physics)*
*Birthday: 1814, Heilbron, Germany*

Julius Robert von Mayer studied medicine at Tubingen but was most interested in physics and investigated the mechanical equivalent of heat and conservation of energy in living as well as inanimate matter. Mayer preceded both Joule and Helmholtz in his deductions but his work was ignored for many years.

# November 26

*Norbert Wiener (Mathematics)*
*Birthday: 1894, Cambridge, Massachusetts, USA*

Norbert Wiener graduated from Harvard with a Ph.D. at age 18, studied mathematical logic under Bertrand Russell at Cambridge, and then joined MIT where he was the stimulus behind the development of computers there. During WWII Wiener was involved as a mathematician in anti-aircraft defense and became interested in computing machines, communications and brain function. He coined the term cybernetics.

*"Edison's greatest invention was that of the industrial research laboratory."*

*Norbert Wiener (U.S. Mathematician) November 26*

# November 27

*Anders Celsius (Astronomy)*
*Birthday: 1701, Uppsala, Sweden*

Anders Celsius was professor of astronomy at Uppsala. He published observations of the aurora borealis and approximate measurements on the relative brightness of stars. In 1742 he devised the temperature scale that bears his name. At first he placed the boiling point of water at zero degrees and freezing point at 100 degrees but in 1743 he reversed these.

# November 28

*John Wesley Hyatt (Invention)*
*Birthday: 1837, Starkey, New York, USA*

John Wesley Hyatt worked as a printer and then opened a factory to manufacture checkers and dominoes. Then, in the early 1860s, he entered a competition for a substitute to ivory for making billiard balls. He invented celluloid, the first synthetic plastic. Its disadvantage was its flammability and it was eventually replaced in 1909 by Bakelite invented by Baekeland.

# November 29

*John Ray (Botany)*
*Birthday: 1628, Black Notley, Essex, England*

John Ray studied at Cambridge. He had a passion for natural history and conceived the idea of describing all living species of plants and animals. He traveled briefly to collect specimens and he published prolifically. His great contribution to science was a system of classification which organized plants, insects and vertebrates. Ray also wrote on other subjects.

# November 30

*Robert Broom (Anthropology)*
*Birthday: 1866, Paisley, Scotland*

Robert Broom qualified in medicine at Glasgow and then spent some years in Australia studying marsupial fossils, and in South Africa studying fossil reptiles. Later he met Raymond Dart and his interest turned to searching for fossil hominids in Africa. In the 1940s he discovered several australopithecines at Sterkfontaine and Swartkrans.

*Australopithecine skull*

# December

*"To myself I seem to have been only like a boy playing on the seashore,
and diverting myself in now and then finding a smoother pebble
or a prettier shell than ordinary, whilst the great ocean
of truth lay all undiscovered before me."*

*Isaac Newton (English physicist) December 25*

*Augusta Ada Byron, Lady Lovelace*

# Introduction to December

## Powerful Stuff

I am often asked why I do what I do in my life and I think the most honest answer is that I do things because I enjoy doing them. The next question is to find out what I enjoy doing most and that is a difficult one to answer! In many ways, I think my career in the science of paleontology, looking for and understanding the meaning of ancient fossils, has been the most rewarding.

I was not a good student at high school and my entry into paleontology was almost certainly a consequence of my famous parents and their work in the same field. It was clear to me that they really enjoyed their work and some of their passion must have rubbed off on me as a youngster.

It is the excitement of the search and the incredible thrill of being the 'first' person to know something that will be of global interest that I find the most stimulating. When you first see a jaw or a set of teeth, partly buried in sands that are millions of years old, you know instantly that a critical new piece of the puzzle has been found. Previous discoveries and theories are either confirmed or open to a new interpretation. This is powerful stuff for me and I love it.

*Richard Leakey*
*May 1997*

# December

| | | | | |
|---|---|---|---|---|
| 1 | Martin Heinrich Klaproth, 1743 - 1817 | | 16 | Johann Wilhelm Ritter, 1776 - 1810 |
| 2 | George Richards Minot, 1885 - 1950 | | | Margaret Mead, 1901 - 1978 |
| | Isabella Karle, 1921 - | | 17 | Emilie du Chatelet, 1706 - 1749 |
| 3 | Carl Koller, 1857 - 1944 | | | Humphry Davy, 1778 - 1829 |
| 4 | Alfred Day Hershey, 1908 - 1997 | | 18 | Joseph John Thomson, 1856 - 1940 |
| 5 | Sheldon Lee Glashow, 1932 - | | 19 | Thomas Andrews, 1813 - 1885 |
| 6 | Joseph Louis Gay-Lussac, 1778 - 1860 | | | Richard E. Frere Leakey, 1944 - |
| | Birgit Zipser, 1943 - | | 20 | Robert J. Van de Graaff, 1901 - 1967 |
| 7 | Theodore Schwann, 1810 - 1882 | | 21 | Herman Joseph Muller, 1890 - 1967 |
| 8 | Eli Whitney, 1765 - 1825 | | 22 | Grote Reber, 1911 - |
| 9 | Carl Wilhelm Scheele, 1742 - 1786 | | 23 | Richard Arkwright, 1732 - 1792 |
| | Grace B. Murray Hopper, 1906 - 1992 | | | Point-contact Transistor, 1947 |
| 10 | Augusta A. B. Lady Lovelace, 1815 - 1852 | | 24 | James Prescott Joule, 1818 - 1889 |
| | Howard Martin Temin, 1934 - 1994 | | 25 | Isaac Newton, 1642 - 1727 |
| 11 | Robert Koch, 1852 - 1910 | | 26 | Mary F. Somerville, 1780 - 1872 |
| | Charles Frederick Cross, 1855 - 1935 | | | Charles Babbage, 1791 - 1871 |
| | Annie Jump Cannon, 1863 - 1941 | | 27 | Johannes Kepler, 1571 - 1630 |
| | Max Born, 1882 - 1970 | | | Louis Pasteur, 1822 - 1895 |
| 12 | Erasmus Darwin, 1731 - 1802 | | 28 | John Bennet Lawes, 1814 - 1899 |
| 13 | Ernst Werner von Siemens, 1816 - 1892 | | | John Von Neumann, 1903 - 1957 |
| 14 | Tycho Brahe, 1546 - 1601 | | 29 | Alexander Parkes, 1813 - 1890 |
| 15 | Antoine Henri Becquerel, 1852 - 1908 | | 30 | Sergei P. Korolev, 1906 - 1966 |
| | | | 31 | Andreas Vesalius, 1514 - 1564 |

# December 1

## Martin Heinrich Klaproth (Chemistry)
### Birthday: 1743, Wernigerode, Germany

Martin Heinrich Klaproth practiced for some years as an apothecary and then became the first professor of chemistry at the new University of Berlin. He was a founder of analytical chemistry and discovered a number of new elements (independently of others who found the same ones) including zirconia, titanium, chromium and tellurium.

---

# December 2

## George Richards Minot (Medicine)
### Birthday: 1885, Boston, Massachusetts, USA

George Richards Minot qualified as a physician at Harvard and became professor of medicine there. He was especially interested in blood disorders and early in his career was convinced that pernicious anemia was a result of some deficiency. He shared a 1934 Nobel Prize for the discovery of (dietary) liver therapy for anemias to replace the use of arsenic and splenectomy.

## Isabella Karle (Crystallography)
### Birthday: 1921, Detroit, Michigan, USA

Isabella Karle (born Isabella Lugoski) received her BS , MA and a Ph.D. in Physical Chemistry from the University of Michigan by the age of 22. In 1943 she married Jerome Karle (June 18). The Karles worked together on developing methods of determining the positions of atoms in crystals directly from photographs. Jerome worked mainly on the theoretical process and Isabella worked on the electron diffraction apparatus they needed to use. Later, Isabella worked on transportation of ions across living cell membranes

# December 3

*Carl Koller (Medicine)*
*Birthday: 1857, Schuttenhoffen, Bohemia*

Carl Koller studied medicine at Vienna and became house surgeon there. Encouraged by his friend Sigmund Freud, he investigated the medical properties of cocaine (an alkaloid from the coca plant) and introduced its use as a local anesthetic in ophthalmology and in nose, throat, and dental surgery. He set himself up as an ophthalmologist in New York.

# December 4

*Alfred Day Hershey (Biology)*
*Birthday: 1908, Owosso, Michigan, USA*

Alfred Day Hershey's first degree was in chemistry and he obtained his Ph.D. with a thesis in bacteriology. He was one of the first bacteriologists to study bacteriophages and, with geneticist Martha Chase, he confirmed Avery's hypothesis that genes are made of DNA, He shared a 1969 Nobel Prize for discovering genetic recombination in viruses.

# December 5

*Quark*

*Sheldon Lee Glashow (Physics)*
*Birthday: 1932, New York, NY, USA*

Sheldon Lee Glashow graduated from Cornell and Harvard. He shared the 1979 Nobel Prize in Physics for his part in developing (1960) a theory explaining electromagnetic force and weak nuclear force. The theory predicted the existence of the weak neutral currents which were discovered in 1973. Later, Glashow predicted the existence of a fourth type of quark (charm). It was found in 1974.

# December 6

*Joseph Louis Gay-Lussac (Chemistry)*
*Birthday: 1778, St Leonard, France*

Joseph Louis Gay-Lussac showed in 1802 that different gases all expand by equal amounts with the same rise in temperature. In 1804 he collected air samples during a 23,000 ft balloon flight. The collected air proved to have the same composition as at sea level. Gay-Lussac also investigated the chemistry of iodine and Prussian blue, and devised an effective way to assay silver.

*Birgit Zipser (Neurobiology)*
*Birthday: 1943, Oldenburg, Germany*

Birgit Zipser received her doctorate in neurobiology from New York's Albert Einstein School of Medicine in 1972. In 1978 she moved to the Cold Spring Harbor Laboratory on Long Island, New York where her research focused on the workings of leech nervous system. Leeches have large neurons and simple behavior patterns which makes them ideal material for discovering the way that networks of neurons interact and communicate. Zipser's work ultimately helps to elucidate the workings of neurons in the human brain and nervous system.

# December 7

*Theodore Schwann (Biology)*
*Birthday: 1810, Neuss, Germany*

Theodore Schwann studied medicine at Bonn. Most of his important work was done in 1834-8 while he was in Berlin at the Museum of Anatomy. While making histological preparations he found and named the delicate myelin sheath surrounding nerve fibers; he discovered pepsin, showed that yeast causes fermentation, and recognized that animal tissues are composed of animal cells.

# December 8

*Eli Whitney (Invention)*
*Birthday: 1765, Westboro, Massachusetts, USA*

Eli Whitney manufactured nails for several years, then studied law at Yale. In Georgia (1792) he learned of the need for a machine to clean green-seed cotton. In 10 days Whitney designed a cotton-gin for which he took out a patent in March 1794. In 1798 he received a contract to manufacture rifles and he introduced the concept of interchangeable replacement parts.

*Cotton*

# December 9

*Carl Wilhelm Scheele (Chemistry)*
*Birthday: 1742, Stralsund, Sweden*

Carl Wilhelm Scheele (whose birthdate is variously referenced as either December 9 or 19) made a number of chemical discoveries. He discovered a number of acids, prepared several poisonous gases, and was involved in the discovery of other chemicals only one of which he is given credit for - Scheele's green (copper arsenate). He discovered oxygen in 1773 but Priestly's similar discovery was published first.

*Grace Brewster Murray Hopper (Computers Science)*
*Birthday: 1906, New York City, USA*

Grace Brewster Murray attended Vassar to study mathematics and physics and received a doctorate from Yale in 1934. As a professor of mathematics she taught at Vassar until 1943 when she joined the US Naval Reserve. She is best known for her work in the Navy where she played a significant part in the design and development of the COBOL computer programming language.

# December 10

*Augusta Ada Byron, Lady Lovelace (Mathematics)*
*Birthday: 1815, London, England*

Augusta Ada Byron, daughter of the poet Byron, was raised by her mother to be a mathematician and scientist. In 1834 she met Charles Babbage (December 26) and became intrigued with his idea for a calculating engine (machine). In 1843 Ada translated from French into English an account of Babbage's plans and added her own very substantial notes. She predicted that such a machine could be used to compose music and calculate, for example, Bernouilli numbers. In 1979 a software language was named Ada in her honor.

*Augusta Ada Byron,*
*Lady Lovelace*

*Howard Martin Temin (Biology)*
*Birthday: 1934, Philadelphia, Pennsylvania, USA*

Howard Martin Temin, a virologist, developed a technique for measuring the growth of a virus and suggested that viruses can alter the genetic code of cells that they attack. In 1970 Temin isolated an enzyme, found in retroviruses, known now as reverse transcriptase, that can transfer viral RNA to cellular DNA. Temin shared the 1975 Nobel Prize in Physiology or Medicine.

# December 11

*Max Born (Physics)*
*Birthday: 1882, Breslau, Silesia*

Max Born studied at Breslau, Heidelberg, Zurich and then Cambridge. He was influenced there by J.J. Thomson (physicist and discoverer of the electron) and was inspired by Niels Bohr to look for a mathematical explanation of the behavior of an electron in an atom. In the 1920's at Gottingen, Born's interest turned to quantum theory. He worked on a statistical explanation  and coined the term quantum mechanics but it was not until 1954 when he was at Edinburgh that his contribution to physics was recognized with the Nobel Prize for his fundamental research in quantum mechanics and his statistical interpretation of the wave function.

# December 11

*Charles Frederick Cross (Chemistry)*
*Birthday: 1855, Brentford, Mddx, England*

Charles Frederick Cross was educated at London, Zurich and Manchester Universities and became an industrial chemist. He specialized in the chemistry and technology of cellulose and lignin. He and his partners in 1892, took out a historic patent which initiated the rayon industry. Wood pulp (cellulose) was dissolved in an aqueous mixture of carbon disulphide and sodium hydroxide and the resulting cellulose xanthate was squirted through fine nozzles into dilute acid to produce thin fibers of viscose (rayon).

*Annie Jump Cannon (Astronomy)*
*Birthday: 1863, Dover, Delaware, USA*

Annie Jump Cannon was an astronomer at Harvard College Observatory from 1896 - 1940. She classified stellar bodies according to their temperature and classified the spectra of all stars from the North Pole to the South Pole. She completed the Henry Draper Catalog which became the basis for modern astronomical stellar spectroscopy and contains 225,320 stars.

*"The seeds of great discovery are constantly floating around us,*
*but they only take root in minds well prepared to receive them."*

*Joseph Henry (American physicist) December 17*

## December 12

*Erasmus Darwin (Medicine)*
*Birthday: 1731, Elston Hall, Nottinghamshire, England*

Erasmus Darwin (grandfather of Charles Darwin) qualified as a physician at Edinburgh and set up in practice in Lichfield. He was a man of originality and intellectual vigor with wide ranging interests. He invented several improvements to a windmill, oil lamp, and steam engine. He was interested in botany and wrote the poem *Botanic Garden*, and *Zoonomia* (Laws of Organic Life).

## December 13

*Ernst Werner von Siemens (Invention)*
*Birthday: 1816, Lenthe, Hanover, Germany*

Ernst Werner von Siemens was imprisoned for dueling and set up a laboratory in his prison cell. Here he experimented on silver and gold plating, and sold the patent rights to his discoveries. In 1846 he used make-and-break circuits to make Wheatstone's dial telegraph self acting. He founded the Siemens and Halske Company to develop his inventions.

## December 14

*Tycho Brahe (Astronomy - Pre-telescopic)*
*Birthday: 1546, Knudstrup, Skaane (Denmark)*

Tycho Brahe studied law at Leipzig and taught himself astronomy. On 11 November, 1572 he saw the nova in Cassiopeia and wrote a careful account of his observations in his *De Nova Stella* (1573). He built at Uraniborg the first modern observatory where he had a wide variety of instruments but no telescopes for they had not yet been invented.

# December 15

*Antoine Henri Becquerel (Physics)*
*Birthday: 1852, Paris, France*

Antoine Henri Becquerel's doctorate thesis was on the absorption of light. He then did investigated to discover if X-rays were produced in fluorescence. He used potassium uranyl sulphate as a source of the fluorescence and discovered that this substance - which contained uranium atoms - in addition to being fluorescent, gave off what he called Becquerel Rays (Marie Curie called them radiation).

# December 16

*Johann Wilhelm Ritter (Chemistry)*
*Birthday: 1776, Samnitz, Silesia*

Johann Wilhelm Ritter worked as a pharmacist, then studied medicine. He wrote on animal galvanism, invented the dry voltaic pile, an electric storage battery, and electroplating with copper sulphate. He discovered ultra-violet radiation by its effect on silver chloride. Ritter's discoveries and theories played a significant part in early electrochemistry.

*Margaret Mead (Anthropologist)*
*Birthday: 1901, Philadelphia, USA*

Margaret Mead was the oldest of five children in an unconventional Philadelphia family. She studied at Barnard College with anthropologist Franz Boas and became a pioneer and innovator in social anthropology. Her studies were conducted in the field in underdeveloped countries and were published in several influential books including *Coming of Age in Samoa*. She concluded that culture is a primary determining factor in adolescent behavior.

*Margaret Mead*

# December 17

*Emilie du Chatelet (Mathematics)*
*Birthday: 1706, Paris, France*

Emilie du Chatelet, born into Paris society, was given an unusual education for the time. Her studies included languages and classics but her real interest was mathematics. She wrote *Institutions de Physique* in 1740 and her translation into French of Newton's (December 25) *Principia* was published posthumously with a preface by Voltaire.

*Humphry Davy (Chemistry)*
*Birthday: 1778, Penzance, Cornwall, England*

Humphry Davy, in 1797, read Lavoisier's Textbook on Chemistry and decided to be a chemist. He discovered nitrous oxide and its unusual properties when breathed. He investigated the effect of electricity on such substances as lime, magnesia, potash and soda. He investigated acids and named chlorine. In 1815 he invented the Davy Safety Lamp for use in the Cornish mines.

# December 18

*Joseph John Thomson (Physics)*
*Birthday: 1856, Cheetham Hill, Manchester, England*

Joseph John Thomson intended at first to be an engineer but became a physicist and worked at the Cavendish Lab in Cambridge. Interested in electromagnetic radiation, Thomson showed in 1897 that cathode rays were particulate and subatomic in size. The particles were called electrons and were believed by Thomson to be a universal component of matter. He won a 1906 Nobel Prize in Physics.

# December 19

*Thomas Andrews (Chemistry)*
*Birthday: 1813, Belfast, Ireland*

Thomas Andrews qualified and practiced as a physician but found time for scientific studies on ozone, heats of chemical combination, and the continuity of liquid and gaseous states. He proved that ozone is oxygen in altered form - not a compound, and that gases can be liquefied by pressure but that for each gas there was a temperature above which increased pressure alone cannot liquefy it.

*Richard Erskine Frere Leakey (Paleoanthropology)*
*Birthday: 1944, Nairobi, Kenya*

Richard Erskine Frere Leakey, an African-born British paleontologist, concentrated his research looking for early humans in northern Kenya and southern Ethiopia. At Koobi Fora his team found more than four hundred hominid fossils that were about 1.8 million years ago. In 1972 Leakey found Skull 1470 the oldest example of *Homo habilis* yet discovered. Leakey believed *Homo habilis* and the Australopithecines to have been contemporaries.

*Australopitecine skull*

# December 20

*Robert Jemison Van de Graaff (Physics)*
*Birthday: 1901, Tuscaloosa, Alabama, USA*

Robert Jemison Van de Graaff graduated from Alabama in 1922 and studied at the Sourbonne and Oxford. He was interested in atomic physics and the mobility of charged ions and is best known for the high voltage, electrostatic generator that he developed in 1931. This Van de Graaff apparatus stores a generated charge inside a hollow metal sphere and can develop a potential of several million volts.

# December 21

*Herman Joseph Muller (Biology)*
*Birthday: 1890, New York, NY, USA*

Herman Joseph Muller studied biology at Columbia to become a geneticist. In 1911 he began work on the genetics of the fruit fly Drosophila and looked for ways of increasing the mutation rate. In 1919 he found that raising the temperature did this. By 1926 he showed that X-rays could increase the mutation rate by as much as 150 times. He was awarded the 1964 Nobel Prize in Physiology or Medicine.

*Drosophila sp.*

# December 22

*Grote Reber (Astronomy)*
*Birthday: 1911, Wheaton, Illinois, USA*

Grote Reber was a radio ham whose special interest was the investigation of cosmic radio waves. By 1937 he had built in his back yard a 30 ft radio wave detector tuned to a wavelength of 30 cm. He found points in the sky from which signals were especially strong but these radio stars were not otherwise visible. One of them was later identified by Baade as being a pair of colliding galaxies.

# December 23

*Richard Arkwright (Invention)*
*Birthday: 1732, Preston, Lancashire, England*

Richard Arkwright was the youngest of seven children and had little formal education He grew up during the time of the Industrial Revolution and became a pioneer in the transformation of the textile industry. He invented and patented a new type of spinning machine in 1769 and although most of his engineering improvements were adaptations of existing machinery his influence as an inventor and in organizing large-scale factory production was significant.

## December 23

*Transistor (Invention)*
*Event: 1947, USA*

On December 23, 1947 Shockley (February 13) and Bardeen's (May 23) electronic, point-contact transistor was demonstrated for the first time. It was used in a demonstration to amplify a human voice. The device was named by John Pierce who chose from two options – amplister or transistor. Ten years after its invention the transistor's price had dropped from $20 to $1.50 per item, and 30 million had been manufactured.

## December 24

*James Prescott Joule (Physics)*
*Birthday: 1818, Salford, Lancashire, England*

James Prescott Joule studied briefly under Dalton learning math, chemistry and the scientific method. He was almost fanatic about measurement. In his teens he published measurements of heat output of electric motors and went on to measure the heat production of every process he could think of. The unit of measurement bearing his name (the joule) is a unit of work (energy.)

## December 25

*Isaac Newton (Mathematics)*
*Birthday: 1642, Woolsthorp, Lincolnshire, England*

Isaac Newton is sometimes said to have been the greatest intellect that ever lived. He attended Cambridge (1660-65) and there, and on his own, investigated optics, dynamics and mathematics in all of which areas his genius advanced the science of the day and developed theories and explanations that remain true today. He was humble about his work and described himself as like a boy playing on the seashore discovering pebbles and shells, while the great ocean of truth lay undiscovered before him.

# December 26

*Mary Fairfax Somerville (Mathematics)*
*Birthday: 1780, Jedburgh, Scotland*

Mary Fairfax Somerville had little formal schooling. She became interested in algebra at age 13 and continued from then to teach herself mathematics and astronomy encouraged later by her second husband William Somerville. In 1825 she started investigating magnetism and presented a paper the following year to the Royal Society. She published *The Mechanism of the Heavens*, *The Connection of the Physical Sciences* (1834), *Physical Geography* (1848), and at age 89, *Molecular and Microscopic Science* (1869).

*Charles Babbage (Mathematics)*
*Birthday: 1791, Southark, London, England*

Charles Babbage invented the occulting lighthouse, and the ophthalmoscope and was an outstanding cryptologist. He is called the grandfather of computer pioneers. In 1822 he invented a mechanical computer, the Difference Engine, for compiling tide tables. His design for a steam powered Analytical Engine was described by his assistant Ada Lovelace, who translated an Italian's report on the engine, as capable of weaving algebraic patterns like a Jacquard loom wove patterns.

# December 27

*Johannes Kepler (Mathematics)*
*Birthday: 1571, Weil, Wurttemburg*

Johannes Kepler studied at Tubingen and was a disciple of Copernican (February 19) astronomy. Fleeing from Catholic Archduke Ferdinand, he joined Tycho Brahe (December 14) in Prague in 1600 and took over Tycho's work when he died. Kepler was fascinated by harmonies in nature, optics, and planetary motion. He developed three laws of planetary motion.

# December 27

*Louis Pasteur (Chemistry)*
*Birthday: 1822, Dole, France*

Louis Pasteur was inspired by Dumas (1800-84) to become a chemist. He investigated the passage of polarized light through crystals and solutions and founded the science of polarimetry. In 1856 he was asked to solve the problem of the souring of wine. He found that living yeasts were involved in wine fermentation and invented pasteurization. He also proposed the germ theory of disease.

*"In the field of observation, chance favors only the prepared mind."*

*Louis Pasteur  December 27, 1822*

# December 28

*John Bennet Lawes (Biology, Agriculture)*
*Birthday: 1814, Rothamsted, Hertfordshire, England*

John Bennet Lawes, a landowner farmer, was the father of the synthetic fertilizer industry. At his estate at Rothamsted he did field trials and careful chemical analysis to learn the mineral needs of a variety of important crops. He established a Trust to maintain Rothamsted as an international research station.

*John Von Neumann (Mathematics)*
*Birthday: 1903, Budapest, Hungary*

John Von Neumann started his career as a chemical engineer but in 1930 he emigrated to the United States to become a professor of mathematics at Princeton University. He is recognized as a pioneer of game theory and computer science.

## December 29

**Alexander Parkes (Chemistry)**
*Birthday: 1813, Birmingham, England*

Alexander Parkes was a metallurgist who silver-plated such delicate objects as spider webs and flowers. From electroplating he turned to molding cellulose nitrate for production of plastic objects. He took out many patents for alloys of copper, nickel, zinc and silver. His name is remembered by the Parkes process for desilvering lead.

*Vostock I*

## December 30

**Sergei Pavlovich Korolev (Engineering)**
*Birthday: 1906, Zhitomir, Ukraine*

Sergei Pavlovich Korolev was an engineer, the chief designer for the Russian Space Program. He was responsible for the design of the Russian spacecraft Vostok I which, on April 12, 1961, carried Yuri Gagarin on man's first orbital space flight. Korolev is described as having intuition in engineering and great creative boldness in problem solving.

## December 31

**Andreas Vesalius (Biology)**
*Birthday: 1514, Brussels, Belgium*

Andreas Vesalius studied medicine and graduated from Padua in 1537. He made two innovations in teaching that gave him great stature as a teacher. He did his own demonstration dissections, and used drawings as teaching aids. He produced *De Corporis Humani Fabrica,* one of the great books of science, which contained exquisite and accurate anatomical drawings - mostly by Calcar.

# ESSAY AUTHORS

~~~◆~~~

*Lars William James Anderson (Plant Physiology)*
*Birthday: April 22, 1945, Pasadena, California, USA*

Lars Anderson studied marine biology and plant physiology at University of California campuses at Irvine as an under graduate, and Santa Barbara for his doctoral research. After two years with the US Environmental Protection Agency in Washington DC, he ran aquatic weed research laboratories in Denver and on the UC Davis campus for the US Department of Agriculture's Agricultural Research Service.

*Thomas A. Cahill (Atmospheric Science and Physics)*
*Birthday: March 4, 1937, Patterson, New Jersey, USA*

Thomas Cahill studied physics at Holy Cross College, B.A. 1959, and earned his doctorate in physics at UC Los Angeles. As a professor at UC Davis he was Director of both the Crocker Nuclear Laboratory and the Institute of Ecology, founded the Air Quality Group and was co-Director of the Crocker Historical and Archaeological Projects. Cahill is now officially retired but continues his research in aerosols and climate, largely in Asia.

*Jacques-Yves Cousteau (Oceanography)*
*Birthday: June 11, 1910, St Andre-de-Cubzac, France*

Jacques-Yves Cousteau entered the French Naval Academy as a young man. During his rehabilitation after an automobile accident, he swam every day, learned to goggle-dive and fell in love with the underwater world. He developed a breathing regulator and other diving gear, and became world famous for his fascination with, and research into, the variety, interdependence and fragility of ocean life. His many books and documentary films helped to popularize the marine environment. Over years of diving and research, he saw the seas and the creatures in them changing for the worse, and founded The Cousteau Society (www.cousteau.org) with a mission to help protect the planet and its oceans.

*Carol Erickson (Developmental Biology)*
*Birthday: September 11, 1949, Wilmington, Delaware, USA*

Carol Erickson is a Professor of Molecular and Cellular Biology at the University of California, Davis, where she has been since 1979. Her research interest since graduate school has focused on how cells move around in the embryo and how they are directed to the correct place for normal development to proceed. In particular, she has determined how a population of cells called neural crest cells migrate during early embryogenesis from the neural tube to give rise to many different derivatives, including neurons, glial cells, pigment cells of the skin and connective tissues of the face (even teeth!).

*Mont Hubbard (Mechanical Engineering)*
*Birthday: June 12, 1943, Fort Smith, Arkansas, USA*

Mont Hubbard joined the University of California, Davis, Mechanical Engineering Department in 1974. His special research interests are the mechanics and biomechanics of sports and the creation of mathematical models of human motion especially in bobsledding, javelin throwing, ski jumping and high jumping. Hubbard's laboratory has developed a virtual reality bobsled driver training simulator that the US Olympic bobsled teams have used to improve their performance. He has also studied and written on the curious behavior of the rattleback.

*Richard Erskine Frere Leakey (Paleoanthropology)*
*Birthday: December 19, 1944, Nairobi, Kenya*

Richard Erskine Frere Leakey, an African-born British paleontologist, concentrated his research looking for early humans in northern Kenya and southern Ethiopia. At Koobi Fora his team found more than four hundred hominid fossils that were about 1.8 million years old. In 1972 Leakey found Skull 1470 the oldest example of Homo habilis yet discovered. Leakey believed Homo habilis and the Australopithecines to have been contemporaries.

*Glenn E. Nedwin (Biochemistry)*
*Birthday: December 9, 1955, Brooklyn, New York, USA*

Glenn Nedwin worked in a tumor biology lab at Brooklyn Jewish Hospital during his summer college years, trying to immunostimulate cancer patients in order to build up their immunity. Later, as a Ph.D. molecular biologist he became a key member of a team of scientists that were the first to clone and express a number of human immunoregulatory cytokines, such as tumor necrosis factors-alpha and beta. Today, these factors have become widely studied in immunology and have been targets for pharmaceutical therapeutic drug development.

*Kelly Stewart (Anthropology)*
*Birthday: May 7, 1951, Los Angeles, California, USA*

Kelly Stewart is a member of the University of California, Davis Anthropology Department. She obtained her doctorate in "Social Development of Wild Mountain Gorillas" in the Zoology Department, University of Cambridge. She worked for five years at the Karisoke Research Center in Rwanda, studying the social behavior, ecology and conservation of mountain gorillas. Her findings have increased understanding of the evolutionary forces that shaped not just gorillas, but all the great apes, including humans.

*Edward Teller (Physics)*
*Birthday: January 15, 1908, Budapest, Hungary*

Edward Teller emigrated to the US in 1935 and became a naturalized citizen. As a theoretical physicist he was part of the 1943-5 team that developed the first fission device in the Manhattan Project (July 16) in Los Alamos, New Mexico. He then worked on hydrogen fusion. In later years he developed the Lawrence Livermore National Laboratory in California to provide a balanced, two-laboratory approach to the problems of nuclear defense.

*Geerat J. Vermeij (Geology)*
*Birthday: September 28, 1946, Sappemeer, Netherlands*

Geerat Vermeij is an evolutionary biologist and paleontologist at the University of California Davis, and a leading modern authority on molluscs. Vermeij is blind. His special expertise is in marine ecology and he has authored several specialized books as well as a biography, "Privileged Hands: A Scientific Life." His research has recently focussed on the time and circumstances of appearance of evolutionary innovations that have a significant economic impact on ecosystems.

*Kenneth L. Verosub (Geology)*
*Birthday: July 10, 1944, New York City, USA*

Kenneth Verosub has been a professor in the University of California, Davis, Geology Department since 1975. He obtained his doctorate from Stanford University and is recognized as a leader in research on the magnetic properties of sediments and soils. His current research interests include the development of new techniques for using magnetic grains as tracers of environmental and paleoclimatic processes. In recent years, his research has taken him to sites ranging from China and Russia to Antarctica and South Africa. He is also recognized locally and statewide as an exceptional teacher and was named the 1997 California Professor of the Year by the Carnegie Foundation for the Advancement of Teaching.

*Richard F. Walters (Computer Science)*
*Birthday: August 30, 1930, Teleajen, Romania*

Richard Walters is now retired but, until 2000, was Professor of Computer Science and of Medical Informatics at the University of California, Davis. His interests and research activities include a focus on distance learning, programming languages, and medical informatics. In 1999 he refined his "Remote Technological Assistance" program which sends, instantaneously, documents from the screen of one networked computer to the screen of another allowing on-screen comments or editing by all collaborating users. Walters named the new program, "Remote Collaboration Tool (RCT)".

# INDEX OF SCIENTISTS & EVENTS

Blackwell, Elizabeth,   February 3, 1821,   Medicine
Bloch, Felix,   October 23, 1905,   Physics
Bohr, Niels Henrik David,   October 7, 1885,   Physics
Boltzmann, Ludvig,   February 20, 1844,   Theoretical Physics
Boole, George,   November 2, 1815,   Mathematics
Borlaug, Norman,   March 25, 1914,   Agriculture
Born, Max,   December 11, 1882,   Physics
Boulton, Matthew,   September 3, 1728,   Engineering, Manufacturing
Bower, Frederick Orpen,   November 4, 1855,   Botany
Boyd, William,   March 4, 1903,   Biochemistry
Boyer, Herbert Wayne,   July 10, 1936,   Genetics
Boyle, Robert,   January 25, 1627,   Physics
Brahe, Tycho,   December 14, 1546,   Astronomy
Bright, Richard,   September 28, 1789,   Medicine
Brinell, Johann August,   November 21, 1849,   Engineering
Broom, Robert,   November 30, 1866,   Anthropology
Buchner, Eduard,   May 20, 1860,   Chemistry
Bunsen, Robert Wilhelm,   March 31, 1811,   Chemistry
Burbank, Luther,   March 7, 1849,   Horticulture
Byrd, Richard Evelyn,   October 25, 1888,   Geography

## C

Cannon, Annie Jump,   December 11, 1863,   Astronomy
Carson, Rachel Louise,   May 27, 1907,   Zoology
Cartwright, Edmund,   April 24, 1743,   Invention
Cavendish, Henry,   October 10, 1731,   Physics
Cayley, Arthur,   August 16, 1821,   Mathematics
Celsius, Anders,   November 27, 1701,   Astronomy
Chadwick, James,   October 20, 1891,   Physics
Challenger, Space Shuttle,   January 28, 1986,   Space Exploration
Chaptal, Jean Antoine Claude,   June 4, 1756,   Chemistry
Chargaff, Erwin,   August 11, 1905,   Biochemistry
Charles, Jacques Alexandre Cesar,   November 12, 1746,   Invention
Chatelet, Emilie du,   December 17, 1706,   Mathematics
Chevreul, Michel Eugene,   August 31, 1786,   Chemistry
Child, Charles Manning,   February 2, 1869,   Zoology
Columbia, Shuttle,   April 12, 1981,   Space Exploration
Condon, Edward Uhler,   March 2, 1902,   Physics
Copernicus, Nicolas,   February 19, 1473,   Astronomy
Cori, Gerty Radnitz,   August 15, 1896,   Biochemistry
Corliss, George Henry,   June 2, 1817,   Engineering
Cormack, Allen McCleod,   February 23, 1924,   Physics
Cousteau, Jacques-Yves,   June 11, 1910,   Oceanography
Cray, Seymour,   September 28, 1925,   Electronics
Crick, Francis Harry Compton,   June 8, 1916,   Biophysics
Cross, Charles Frederick,   December 11, 1855,   Chemistry
Curie, Marya (Marie) Sklodowska,   November 7, 1867,   Physics, Radiochemistry
Curie, Pierre,   May 15, 1859,   Physics
Curtis, William,   January 11, 1746,   Botany
Cushing, Harvey,   April 8, 1869,   Medicine
Cuvier, George Leopold,   August 23, 1769,   Natural History

## D

D'Alembert, Jean le Rond,   November 16, 1717,   Mathematics
da Vinci, Leonardo,   April 15, 1452,   Invention
Daguerre, Louis Jacques Mande,   November 18, 1787,   Invention
Dalton, John,   September 5, 1766,   Chemistry
Daly, Marie Maynard,   April 16, 1921,   Biochemistry
Darwin, Charles Robert,   February 12, 1809,   Natural History
Darwin, Erasmus,   December 12, 1731,   Medicine
Davy, Humphry,   December 17, 1778,   Chemistry

de Havilland, Geoffrey,   July 27, 1882,   Engineering, Aeronautics
De Laval, Carl Gustaf Patrik,   May 9, 1845,   Invention
Delambre, Jean Baptiste Joseph,   September 19, 1749,   Mathematics
Dewar, James,   September 20, 1842,   Chemistry
Douglass, Andrew Ellicott ,   July 5, 1857,   Dendrochronology
Drake, Edwin Laurentine,   March 29, 1819,   Oil Drilling
Draper, John William,   May 5, 1811,   Chemistry
Dumas, Jean Baptiste Andre,   July 14, 1800,   Organic Chemistry


**E**

Eckert, John Presper,   April 9, 1919,   Physics
Edinger, Johanna Gabrielle,   November 13, 1897,   Paleontologist
Edison, Thomas Alva,   February 11, 1847,   Invention
Ehrlich, Paul,   March 14, 1854,   Medicine
Ehrlich, Paul Ralph,   May 29, 1932,   Population Biology
Einstein, Albert,   March 14, 1879,   Theoretical Physics
Elion, Gertrude B.,   January 23, 1918,   Biochemistry
Ericsson, John,   July 31, 1803,   Invention
Erlanger, Joseph,   January 5, 1874,   Biology
Essen, Louis,   September 6, 1908,   Physics
Euler, Leonhard,   April 15, 1707,   Mathematics
Evans, Alice,   January 29, 1881,   Biology
Explorer I,   January 31, 1958,   Space Exploration


**F**

Fahrenheit, Gabriel,   May 14, 1686,   Physics
Faraday, Michael,   September 22, 1791,   Physics
Fermat, Pierre de,   August 17, 1601,   Mathematics
Fermi, Enrico,   September 29, 1901,   Nuclear Physics
Feynman, Richard,   May 11, 1918,   Physics
Fibiger, Johannes Andreas Grib,   April 23, 1867,   Medicine
Fischer, Herman Emil,   October 9, 1852,   Chemistry
Flamsteed, John,   August 19, 1646,   Astronomy
Fleming, Alexander,   August 6, 1881,   Medicine
Flugge-Lotz, Irmgard,   July 16, 1903,   Engineering
Fossey, Dian,   January 16, 1932,   Anthropology
Foucault, Leon,   September 18, 1819,   Physics
Fourcroy, Antoine Francoise de,   June 15, 1755,   Chemistry
Frankland, Edward,   January 18, 1825,   Chemistry
Franklin, Benjamin,   January 17, 1706,   Physics
Franklin, Rosalind Elsie,   July 25, 1920,   X-Ray Crystallography
Fraunhofer, Joseph von,   March 6, 1787,   Optics
Fredholm, Eric Ivar,   April 7, 1866,   Mathematics
Freud, Sigmund,   May 6, 1856,   Psychology
Freyssinet, Marie Eugene Leon,   July 13, 1879,   Civil Engineering
Friend, Charlotte,   March 11, 1921,   Microbiology
Friendship 7,   February 20, 1962,   Space Exploration
Friese-Greene, William,   September 7, 1855,   Invention
Frisch, Otto Robert,   October 1, 1904,   Physics
Fuchs, Leonhart,   January 17, 1501,   Botany


**G**

Gabor, Dennis,   June 5, 1900,   Invention
Galilei, Galileo,   February 18, 1564,   Physics
Galvani, Luigi,   September 9, 1737,   Biology
Gates, William,   October 28, 1955,   Computers
Gauss, Karl Friederick,   April, 30, 1777,   Mathematics, Physics
Gay-Lussac, Joseph Louis,   December 6, 1778,   Chemistry
Geiger, Hans Wilhelm,   September 30, 1882,   Nuclear Physics
Germain, Sophie,   April 1, 1776,   Mathematics

Gilbert, Walter,   March 21, 1932,   Molecular Biology
Gilbert, William,   May 24, 1544,   Physics
Glaser, Donald Arthur,   September 21, 1926,   Physics
Glashow, Sheldon Lee,   December 5, 1932,   Physics
Goddard, Robert Hutchings,   October 5, 1882,   Physics
Goeppert-Mayer, Maria,   June 28, 1906,   Mathematical Physics
Golgi, Camillo,   July 7, 1844,   Histology
Goodall, Jane,   April 3, 1934,   Ethology
Gould, John,   September 14, 1804,   Ornithology
Granville, Evelyn Boyd,   May 1, 1924,   Mathematics
Grimaldi, Francesco Maria,   April 2, 1618,   Physics

## H

Hahn, Otto,   March 8, 1879,   Chemistry
Haldane, John Burdon Sanderson,   November 5, 1892,   Genetics
Haldane, John Scott,   May 3, 1860,   Physiology
Hale, George Ellery,   June 29, 1868,   Astronomy
Hales, Stephen,   September 17, 1677,   Biology
Halley, Edmund,   October 29, 1656,   Astronomy
Hamilton, Alice,   February 27, 1869,   Medicine
Hamilton, William Rowan,   August 4, 1805,   Physics, Mathematics
Harden, Arthur,   October 12, 1865,   Biochemistry
Harvey, William,   April 1, 1578,   Medicine
Hawking, Stephen William,   January 8, 1942,   Physics
Haworth, Walter N.,   March 19, 1883,   Chemistry
Hazen, Elizabeth Lee,   August 24, 1885,   Medicine
Heilbron, Ian Morris,   November 6, 1886,   Chemistry
Heinkel, Ernst,   July 24, 1888,   Engineering
Helmholtz, Herman,   August 31, 1821,   Physiology
Hensen, Victor,   February 10, 1835,   Biology
Herschel, Caroline,   March 16, 1750,   Mathematics
Herschel, William,   November 15, 1738,   Astronomy
Hershey, Alfred Day,   December 4, 1908,   Biology
Hertz, Heinrich Rudolph,   February 22, 1857,   Physics
Hess, Victor Franz,   June 24, 1883,   Physics
Hillary, Edmund Percival,   July, 20, 1919,   Mountaineering
Hitzig, Eduard,   February 6, 1838,   Medicine
Hodgkin, Dorothy Crowfoot,   May 12, 1910,   Physical Chemistry
Hofstadter, Robert,   February 5, 1915,   Physics
Holmes, Oliver Wendell,   August 29, 1809,   Medicine
Hooke, Robert,   July 18, 1635,   Physics
Hooker, Joseph Dalton,   June 30, 1817,   Botany
Hopper, Grace Brewster Murray,   December 9, 1906,   Computers
Houdry, Eugene Jules,   April 18, 1892,   Chemical Engineering
Hounsfield, Godfrey Newbold,   August 28, 1919,   Electrical Engineering
Hrdy, Sarah Blaffer,   July 11, 1946,   Anthropology
Hubble, Edwin Powell,   November 20, 1889,   Astronomy
Hutton, James,   June 3, 1726,   Geology
Huxley, Thomas Henry,   May 4, 1825,   Zoology
Huygens, Christiaan,   April 14, 1629,   Mathematics
Hyatt, John Wesley,   November 28, 1837,   Invention

## J

Jackson, Shirley Ann,   August 5, 1946,   Physics
Jemison, Mae Carol,   October 17, 1956,   Medicine
Jenner, Edward,   May 17, 1749,   Medicine
Joliot-Curie, Irene,   September 12, 1897,   Radiochemistry
Josephson, Brian David,   January 4, 1940,   Physics
Joule, James Prescott,   December 24, 1818,   Physics
Jung, Carl Gustav,   July 26, 1875,   Psychology
Kane, Robert John,   September 24, 1809,   Chemistry

Karl, Jerome,   June 18, 1918,   Physical Chemistry
Karle, Isabella,   December 2, 1921,   Crystallography
Katz, Bernard,   March 26, 1911,   Physiology
Kelly, William,   August 21, 1811,   Invention
Kendrew, John Cowdery,   March 24, 1917,   Physics and Biochemistry
Kepler, Johannes,   December 27, 1571,   Mathematics
Khorana, Har Gobind,   January 9, 1922,   Biochemistry
Klaproth, Martin Heinrich,   December 1, 1743,   Chemistry
Koch, Robert,   December 11, 1852,   Bacteriology
Kohlrausch, Friedrich Wilhelm,   October 14, 1840,   Physics
Koller, Carl,   December 3, 1857,   Medicine
Kolliker, Rudolph Albert Von,   July, 6, 1817,   Biology
Kopp, Hermann Franz Moritz,   October 30, 1817,   Chemistry
Kornberg, Arthur,   March 3, 1918,   Biochemistry
Korolev, Sergei Pavlovich,   December 30, 1906,   Engineering
Krebs, Edwin G.,   June 6, 1918,   Biochemistry
Krebs, Hans Adolph,   August 25, 1900,   Biochemistry
Kuhne, Wilhelm Friedrich,   March 28, 1837,   Physiology

# L

Lamarck, Jean Baptiste,   August 1, 1744,   Natural History
Lancisi, Giovanni Maria,   October 26, 1654,   Medicine
Land, Edwin,   May 7, 1909,   Chemistry
Langevin, Paul,   January 23, 1872,   Physics
Langmuir, Irving,   January 31, 1881,   Physical Chemistry
Latour, Cagniard de,   May 31, 1777,   Physics
Laveran, Charles Louis Alphonse,   June 18, 1845,   Medicine
Lavoisier, Antoine Laurent,   August 26, 1743,   Chemistry
Lawes, John Bennet,   December 28, 1814,   Biology, Agriculture
Le Verrier, Urbain Jean Joseph,   March 11, 1811,   Astronomy
Leakey, Louis Seymour Bazett,   August 7, 1903,   Paleoanthropology
Leakey, Mary Douglas,   February 6, 1913,   Paleoanthropology
Leakey, Richard Erskine Frere,   December 19, 1944,   Paleoanthropology
Leeuwenhoek, Antony van,   October 24, 1632,   Biology, Invention
Levene, Phoebus Aaron Theodor,   February 25, 1869,   Biochemistry
Levi-Montalcini, Rita,   April 22, 1909,   Embryology
Levinstein, Ivan,   July 4, 1845,   Chemistry
Liebig, Justus Von,   May 12, 1803,   Chemistry
Lindbergh, Charles A.,   February 4, 1902,   Aviation
Lindemann,  Carl Louis Ferdinand,   April 12, 1852,   Mathematics
Linnaeus, Carl,   May 23, 1707,   Taxonomy
Lister, Joseph,   April 5, 1827,   Medicine
Lodge, Oliver Joseph,   June 12, 1851,   Physics
Lonsdale, Kathleen Yardley,   January 28, 1903,   Crystallography
Lovelace, Augusta Ada ,   December 10, 1815,   Mathematics
Lumiere, Auguste,   October 19, 1862,   Invention
Lyell, Charles,   November 14, 1797,   Geology

# M

Macewen, William,   June 22, 1848,   Medicine, Surgery
Macquer, Pierre Joseph,   October 9, 1718,   Chemistry
Magendie, Francoise,   October 6, 1783,   Medicine
Maiman, Theodore Harold,   July 11, 1927,   Physics
Malpighi, Marcello,   March 10, 1628,   Medicine
Malthus, Thomas Robert,   February 17, 1766,   Economics
Manson, Patrick,   October 3, 1844,   Medicine
Marconi, Guglielmo,   April 25, 1874,   Invention
Marina 2,   August 27, 1962,   Space Exploration
Mariner 10,   November 3, 1973,   Space Exploration
Marsh, James,   September 2, 1794,   Chemistry

Martin, Archer John Porter,   March 1, 1910,   Biochemistry
Maupertius, Pierre Louis Moreau de,   July 17, 1698,   Mathematics, Geodesy
Maury, Matthew Fontaine,   January 14, 1806,   Oceanography
Maxwell, James Clerk,   June 13, 1831,   Physics
Mayer, Julius Robert von,   November 25, 1814,   Physics
McClintock, Barbara,   June 16, 1902,   Genetics
McCormick, Cyrus Hall,   February 15, 1809,   Engineering
Me(t)chnikov, Ilya (Elie),   May 16, 1845,   Zoology
Mead, Margaret,   December 16, 1901,   Anthropology
Medawar, Peter Brian,   February 28, 1915,   Biology
Meitner, Lise,   November 7, 1878,   Nuclear Physics
Mendel, Johann Gregor,   July 22, 1822,   Genetics
Mendeleeff, Dmitri Ivanovich,   January 27, 1834,   Chemistry
Menzel, Donald Howard,   April 11, 1901,   Astronomy
Mercator, Gerardus,   March 5, 1512,   Cartography
Mercer, John,   February 21, 1791,   Chemistry
Midgley, Thomas,   May 18, 1889,   Chemistry
Millikan, Robert Andrews,   March 22, 1868,   Physics
Minot, George Richards,   December 2, 1885,   Medicine
Mobius, August Ferdinand,   November 17, 1790,   Astronomy
Monge, Gaspard,   May 10, 1746,   Mathematics
Montgolfier, Jacques-Etienne,   January 7, 1745,   Invention
Moore, Stanford,   September 4, 1913,   Biochemistry
Morgan, Ann Haven,   May 6, 1882,   Zoology
Morgan, Thomas Hunt,   September 25, 1866,   Genetics
Morris, Desmond John,   January 24, 1928,   Zoology, Ethology
Morse, Samuel Finley Breese,   April 27, 1791,   Invention
Moseley, Henry Gwyn Jeffreys,   November 23, 1887,   Chemistry
Mount Vesuvius,   August 24, 79,   Disaster
Muir, John,   April 21, 1838,   Naturalist
Muller, Herman Joseph,   December 21, 1890,   Biology

## N

Naudin, Charles,   August 14, 1815,   Botany
Neel, Louis Eugene Felix,   November 22, 1904,   Physics
Nernst, Herman Walther,   June 25, 1864,   Chemistry
Newcomb, Simon,   March 12, 1835,   Mathematics, Astronomy
Newcomen, Thomas,   February 24, 1663,   Engineering
Newton, Isaac,   December 25, 1642,   Mathematics
Nieuwland, Julius Arthur,   February 14, 1878,   Chemistry
Nightingale, Florence,   May 12, 1820,   Mathematics
Nobel, Alfred Bernhard,   October 21, 1833,   Invention
Noether, Amelie Emmy,   March 23, 1882,   Mathematics

## O

Ohm, Georg Simon,   March 16, 1789,   Physics
Oort, Jan Hendrik,   April 28, 1900,   Astronomy
Oppenheimer, Robert,   April 22, 1904,   Physics
Otis, Elisha Graves,   August 3, 1811,   Invention

## P

Palade, George Emil,   November 19, 1912,   Biology
Papin, Denis,   August 22, 1647,   Engineering
Parkes, Alexander,   December 29, 1813,   Chemistry
Partington, James Riddick,   June 20, 1886,   Chemistry, History of Science
Pascal, Blaise,   June 19, 1623,   Mathematics
Pasteur, Louis,   December 27, 1822,   Chemistry
Pathfinder,   July 4, 1997,   Space Exploration
Pauling, Linus,   February 28, 1901,   Chemistry
Pavlov, Ivan Petrovich,   September 26, 1849,   Physiology

Perutz, Max Ferdinand,   May 19, 1914,   Molecular Biology
Picard, Charles Emil,   July 24, 1856,   Mathematics
Picard, Jean,   July 21, 1620,   Astronomy
Planck, Max Karl Ernst Ludvig,   April 23, 1858,   Physics
Playfair, Lyon,   May 21, 1818,   Chemistry
Poincare, Jules Henri,   April 29, 1854,   Mathematics
Porter, Keith Roberts,   June 11, 1912,   Cell Biology
Poulsen, Valdemar,   September 23, 1869,   Invention
Prelog, Vladimir,   July 23, 1906,   Chemistry
Priestly, Joseph,   March 13, 1733,   Chemistry
Pritchard, Charles,   February 29, 1808,   Astronomy
Purcell, Edward Mills,   August 30, 1912,   Physics

## R

Ramey, Estelle,   August 23, 1917,   Physiology
Ray, Dixie Lee,   September 3, 1914,   Marine Biology
Ray, John,   November 29, 1628,   Botany
Reaumur, Rene Antoine Ferchault de,   February 28, 1683,   Physical Science
Reber, Grote,   December 22, 1911,   Astronomy
Reed, Walter,   September 13, 1851,   Medicine
Richter, Burton,   March 22, 1931,   Physics
Ricketts, Howard Taylor,   February 9, 1871,   Medicine
Ride, Sally Kristin,   May 26, 1951,   Physics
Ritter, Johann Wilhelm,   December 16, 1776,   Chemistry
Roentgen, Wilhelm Conrad,   March 27, 1845,   Physics
Ross, Ronald,   May 13, 1857,   Medicine
Russell, Elizabeth Shull,   May 1, 1913,   Genetics
Russell, Frederick Stratten,   November 3, 1897,   Biology
Rutherford, Ernest,   August 30, 1871,   Nuclear Physics
Ryle, Martin,   September 27, 1918,   Physics

## S

Sabin, Florence Rena,   November 9, 1871,   Medicine
Sachs, Ferdinand Gustav Julius von,   October, 2, 1832,   Botany
Sagan, Carl Edward,   November 9, 1934,   Astrophysics
Salk, Jonas Edward,   October 28, 1914,   Microbiology
Salyut 4,   February 2, 1977,   Space Exploration
Scheele, Carl Wilhelm,   December 9, 1742,   Chemistry
Schmidt, Bernhard Voldemar,   March 30, 1879,   Engineering
Schonbein, Christian Friedrich,   October 18, 1799,   Chemistry
Schrodinger, Erwin,   August 12, 1887,   Physics
Schwann, Theodore,   December 7, 1810,   Biology
Seaborg, Glen Theodore,   April 19, 1912,   Chemistry
Segre, Emilio,   February 1, 1905,   Nuclear Physics
Seguin, Marc,   April 20, 1786,   Civil Engineering
Semmelweiss, Ignaz Philipp,   July 1, 1818,   Medicine
Sessions, Kate,   November 8, 1857,   Horticulture
Shockley, William Bradford,   February 13, 1910,   Physics
Shoemaker, Eugene Merle,   April 28, 1928,   Planetary Geology
Sidgwick, Nevil Vincent,   May 8, 1873,   Chemistry
Siemens, Charles William,   April 4, 1823,   Engineering
Siemens, Ernst Werner von,   December 13, 1816,   Invention
Simpson, James Young,   June 7, 1811,   Medicine
Simpson, Joanne,   March 23, 1923,   Meteorology
Sinclair, Clive Marles,   July 30, 1940,   Invention
Singer, Isaac Merrit,   October 27, 1811,   Invention
Snowstorm, First Artificial,   November 13, 1946,   Invention
Somerville, Mary Fairfax,   December 26, 1780,   Mathematics
Sorensen, Soren Peter Lauritz,   January 9, 1868,   Biochemistry
Soyuz 14,   July 3, 1974,   Space Exploration

Spallanzani, Lazarro,   January 12, 1729,   Natural History
Spemann, Hans,   June 27, 1869,   Zoology, Embryology
Sperry, Roger Wolcott,   August 20, 1913,   Zoology
Sputnik I,   October 4, 1957,   Space Exploration
Starling, Ernest Henry,   April 17, 1866,   Physiology
Steno, Nicolaus,   January 1, 1638,   Biology
Stephenson, George,   June 9, 1781,   Engineering
Stephenson, Robert,   October 16, 1803,   Engineering
Stewart, Alice,   October 4, 1906,   Medicine
Stirling, Robert,   October 25, 1790,   Engineering, Invention
Sturgeon, William,   May 22, 1783,   Invention
Surveyor 7,   January 6, 1968,   Space Exploration
Swan, Joseph Wilson,   October 31, 1828,   Invention
Sydenham, Thomas,   September 10, 1624,   Medicine
Szent-Gyorgyi, Albert von,   September 16, 1893,   Biochemistry

## T

Taylor, Brook,   August 18, 1685,   Mathematics
Telford, Thomas,   August 9, 1757,   Civil Engineering
Teller, Edward,   January 15, 1908,   Physics
Temin, Howard Martin,   December 10, 1934,   Biology
Tesla, Nikola,   July 9, 1856,   Electrical Engineering
Theiler, Max,   January 30, 1899,   Medicine
Thomson, Joseph John,   December 18, 1856,   Physics
Thomson, William (Lord Kelvin),   June 26, 1824,   Physics
Tinbergen, Nicolaas,   April 15, 1907,   Ethology
Tiselius, Arne Wilhelm Kaurin,   August 10, 1902,   Biochemistry
Titanic, RMS,   April 14, 1912,   Disaster
Tombaugh, Clyde William,   February 4, 1906,   Astronomy
Tomonaga, Sin-Itiro,   March 31, 1906,   Physics
Torricelli, Evangelista,   October 15, 1608,   Mathematics, Physics
Townes, Charles Hard,   July 28, 1915,   Physics
Transistor, Point-contact,   December 23, 1947,   Invention
Trevithick, Richard,   April 13, 1771,   Engineering
Trinity Test,   July 16, 1945,   Nuclear Physics
Turing, Alan Mathison,   June 23, 1912,   Mathematics
Tyndall, John,   August 2, 1820,   Physics

## V

Van de Graaff, Robert Jemison,   December 20, 1901,   Physics
Vaucanson, Jacques de,   January 24, 1709,   Invention
Vesalius, Andreas,   December 31, 1514,   Biology
Vine, Frederick John,   June 17, 1939,   Geology
Virchow, Rudolph Ludwig Carl,   October 13, 1821,   Biology
Volta, Alessandro,   February 18, 1745,   Invention
Von Euler, Ulf Svante,   February 7, 1905,   Biology
Von Neumann, John,   December 28, 1903,   Mathematics
Voskshod 2,   March 18, 1965,   Space Exploration
Vostock I,   April 12, 1961,   Space Exploration

## W

Wallace, Alfred Russel,   January 8, 1823,   Natural History
Warburg, Otto Heinrich,   October 8, 1883,   Biochemistry
Watson, James Dewey,   April 6, 1928,   Molecular Biology
Watt, James,   January 19, 1736,   Engineering
Wedgewood, Josiah,   July 12, 1730,   Ceramics
Wegener, Alfred Lothar,   November 1, 1880,   Geology
Wells, Horace,   January 21, 1815,   Medicine
Whinefield, John Rex,   February 16, 1901,   Chemistry
Whitehead, Robert,   January 3, 1823,   Engineering

Whitney, Eli,   December 8, 1765,   Invention
Whittle, Frank,   June 1, 1907,   Engineering
Wien, Wilhelm,   January 13, 1864,   Physics
Wiener, Norbert,   November 26, 1894,   Mathematics
Wilcke, Johan Carl,   September 6, 1732,   Physics
Williamson, Alexander,   May 1, 1824,   Chemistry
Williamson, William Crawford,   November 24, 1816,   Biology
Wilson, Robert Woodrow,   January 10, 1936,   Astronomy
Withering, William,   March 17, 1741,   Medicine, Botany
Wolf, Maximilian,   June 21, 1863 - 1932, Astronomy
Wong-Staal, Flossie,   August 27, 1946,   Genetics
Woodward, Robert Burns,   April 10, 1917,   Organic Chemistry
Wren, Christopher,   October 20, 1632,   Architecture, Mathematics
Wright, Orville,   August 19, 1871,   Invention
Wright, Wilbur,   April 16, 1867,   Invention
Wu, Chien-Shiun,   May 29 1912,   Physics

# Y

Yalow, Rosalyn Sussman,   July 19, 1921,   Medical Physics
Young, Arthur ,   September 11, 1741,   Agriculture
Young, Grace Chisholm,   March 15, 1868,   Mathematics
Young, John Zachary,   March 18, 1907,   Biology

# Z

Zeeman, Pieter,   May 25, 1865,   Physics
Zipser, Birgit,   December 6, 1943,   Neurobiology

*If we have omitted a scientist who you think should have been included, please let us know the person's full name, birth date (day, month and year), year of death if appropriate, town and country of birth, and something about what he or she did as a scientist.*

# INDEX OF WOMAN SCIENTISTS

## K

Karle, Isabella,   December 2, 1921,   Crystallography

## L

Leakey, Mary Douglas,   February 6, 1913,   Paleoanthropology
Levi-Montalcini, Rita,   April 22, 1909,   Embryology
Lonsdale, Kathleen Yardley,   January 28, 1903,   Crystallography
Lovelace, Augusta Ada Byron,   December 10, 1815,   Mathematics

## M

McClintock, Barbara,   June 16, 1902,   Genetics
Mead, Margaret,   December 16, 1901,   Anthropology
Meitner, Lise,   November 7, 1878,   Nuclear Physics
Morgan, Ann Haven,   May 6, 1882,   Zoology

## N

Nightingale, Florence,   May 12, 1820,   Mathematics
Noether, Amelie Emmy,   March 23, 1882,   Mathematics

## R

Ramey, Estelle,   August 23, 1917,   Physiology
Ray, Dixie Lee,   September 3, 1914,   Marine Biology
Ride, Sally Kristin,   May 26, 1951,   Physics
Russell, Elizabeth Shull,   May 1, 1913,   Genetics

## S

Sabin, Florence Rena,   November 9, 1871,   Medicine
Sessions, Kate,   November 8, 1857,   Horticulture
Simpson, Joanne,   March 23, 1923,   Meteorology
Somerville, Mary Fairfax,   December 26, 1780,   Mathematics
Stewart, Alice,   October 4, 1906,   Medicine

## W

Wong-Staal, Flossie,   August 27, 1946,   Genetics
Wu, Chien-Shiun,   May 29, 1912,   Physics

## Y

Yalow, Rosalyn Sussman,   July 19, 1921,   Medical Physics
Young, Grace Chisholm,   March 15, 1868,   Mathematics

## Z

Zipser, Birgit,   December 6, 1943,   Neurobiology

# SUBJECT INDEX

Biology, Vesalius, December 31
Biology, Virchow, October 13
Biology, Von Euler, February 7
Biology, Williamson, November 24
Biology, Young, March 18

## Biophysics
Biophysics, Molecular Biology, Crick, June 8

## Botany
Botany, Bower, November 4
Botany, Curtis, January 11
Botany, Fuchs, January 17
Botany, Hooker, June 30
Botany, Naudin, August 14
Botany, Ray, November 29
Botany, Sachs, October 2

## Cartography
Cartography, Mercator, March 5

## Cell Biology
Cell Biology, Porter, June 11

## Ceramics
Ceramics, Wedgewood, July 12

## Chemical Engineering
Chemical Engineering, Houdry, April 18

## Chemistry
Chemistry, Andrews, December 19
Chemistry, Aston, September 1
Chemistry, Auerbach, May 14
Chemistry, Babcock, October 22
Chemistry, Bartlett, September 15
Chemistry, Barton, September 8
Chemistry, Bergius, October 11
Chemistry, Bergman, March 20
Chemistry, Buchner, May 20
Chemistry, Bunsen, March 31
Chemistry, Chaptal, June 4
Chemistry, Chevreul, August 31
Chemistry, Cross, December 11
Chemistry, Dalton, September 5
Chemistry, Davy, December 17
Chemistry, Dewar, September 20
Chemistry, Draper, May 5
Chemistry, Fischer, October 9
Chemistry, Fourcroy, June 15
Chemistry, Frankland, January 18

Chemistry, Gay-Lussac, December 6
Chemistry, Hahn, March 8
Chemistry, Haworth, March 19
Chemistry, Heilbron, November 6
Chemistry, Kane, September 24
Chemistry, Klaproth, December 1
Chemistry, Kopp, October 30
Chemistry, Land, May 7
Chemistry, Langmuir, January 31
Chemistry, Lavoisier, August 26
Chemistry, Levinstein, July 4
Chemistry, Liebig, May 12
Chemistry, Macquer, October 9
Chemistry, Marsh, September 2
Chemistry, Mendeleeff, January 27
Chemistry, Mercer, February 21
Chemistry, Midgley, May 18
Chemistry, Moseley, November 23
Chemistry, Nernst, June 25
Chemistry, Nieuwland, February 14
Chemistry, Parkes, December 29
Chemistry, History of Science, Partington, June 20
Chemistry, Pasteur, December 27
Chemistry, Pauling, February 28
Chemistry, Playfair, May 21
Chemistry, Prelog, July 23
Chemistry, Priestly, March 13
Chemistry, Ritter, December 16
Chemistry, Scheele, December 9
Chemistry, Schonbein, October 18
Chemistry, Seaborg, April 19
Chemistry, Sidgwick, May 8
Chemistry, Whinefield, February 16
Chemistry, Williamson, May 1

## Civil Engineering
Civil Engineering, Freyssinet, July 13
Civil Engineering, Seguin, April 20
Civil Engineering, Telford, August 9

## Computers
Computers, Gates, October 28
Computers, Hopper, December 9

## Crystallography
Crystallography, Karle, December 2
Crystallography, Lonsdale, January 28

## Dendrochronology
Dendrochronology, Douglass, July 5

## Disaster
Disaster, Mount Vesuvius, August 24
Disaster, RMS Titanic, April 14

Physics, Josephson, January 4
Physics, Joule, December 24
Physics, Biochemistry, Kendrew, March 24
Physics, Kohlrausch, October 14
Physics, Langevin, January 23
Physics, Latour, May 31
Physics, Lodge, June 12
Physics, Maiman, July 11
Physics, Maxwell, June 13
Physics, Mayer, November 25
Physics, Millikan, March 22
Physics, Neel, November 22
Physics, Ohm, March 16
Physics, Oppenheimer, April 22
Physics, Planck, April 23
Physics, Purcell, August 30
Physics, Richter, March 22
Physics, Ride, May 26
Physics, Roentgen, March 27
Physics, Ryle, September 27
Physics, Schrodinger, August 12
Physics, Shockley, February 13
Physics, Teller, January 15
Physics, Thomson, June 26
Physics, Thomson, December 18
Physics, Tomonaga, March 31
Physics, Mathematics, Torricelli, October 15
Physics, Townes, July 28
Physics, Tyndall, August 2
Physics, Van de Graaff, December 20
Physics, Wien, January 13
Physics, Wilcke, September 6
Physics, Wu, May 29
Physics, Zeeman, May 25

## Physiology

Physiology, Bayliss, May 2
Physiology, Biology, Medicine, Bernard, July 12
Physiology, Haldane, May 3
Physiology, Helmholtz, August 31
Physiology, Katz, March 26
Physiology, Kuhne, March 28
Physiology, Pavlov, September 26
Physiology, Ramey, August 23
Physiology, Starling, April 17

## Planetary Geology

Planetary Geology, Shoemaker, April 28

## Population Biology

Population Biology, Ehrlich, May 29

## Psychology

Psychology, Binet, July 8
Psychology, Freud, May 6
Psychology, Jung, July 26

## Radiochemistry

Radiochemistry, Joliot-Curie, September 12

## Science, Invention, Education of the Deaf

Science, Invention, Education of the Deaf, Bell, March 3

## Space Exploration

Space Exploration, Aldrin, January 20
Space Exploration, Apollo 11, July 20
Space Exploration, Aeronautics, Armstrong, August 5
Space Exploration, Challenger, January 28
Space Exploration, Columbia, April 12
Space Exploration, Explorer I, January 31
Space Exploration, Friendship 7, February 20
Space Exploration, Mariner 2, August 27
Space Exploration, Mariner 10, November 3
Space Exploration, Pathfinder, July 4
Space Exploration, Salyut 4, February 2
Space Exploration, Soyuz 14, July,3
Space Exploration, Sputnik I, October 4
Space Exploration, Surveyor 7, January 6
Space Exploration, Voskshod 2, March 18
Space Exploration, Vostock I, April 12

## Surgery

Surgery, Lister, April 5
Surgery, Barnard, November 8
Surgery, Medicine, Macewen, June 22

## Taxonomy

Taxonomy, Artedi, February 27
Taxonomy, Linnaeus, May 23

## Theoretical Physics

Theoretical Physics, Einstein, March 14

## X-ray Crystallography

X-ray Crystallography, Franklin, July 25

## Zoology

Zoology, Natural History, Geology, Agassiz, May 28
Zoology, Andrews, January 26
Zoology, Carson, May 27
Zoology, Child, February 2
Zoology, Huxley, May 4
Zoology, Me(t)chnikov, May 16
Zoology, Morgan, May 6
Zoology, Ethology, Morris, January 24
Zoology, Embryology, Spemann, June 27
Zoology, Sperry, August 20